T0391065

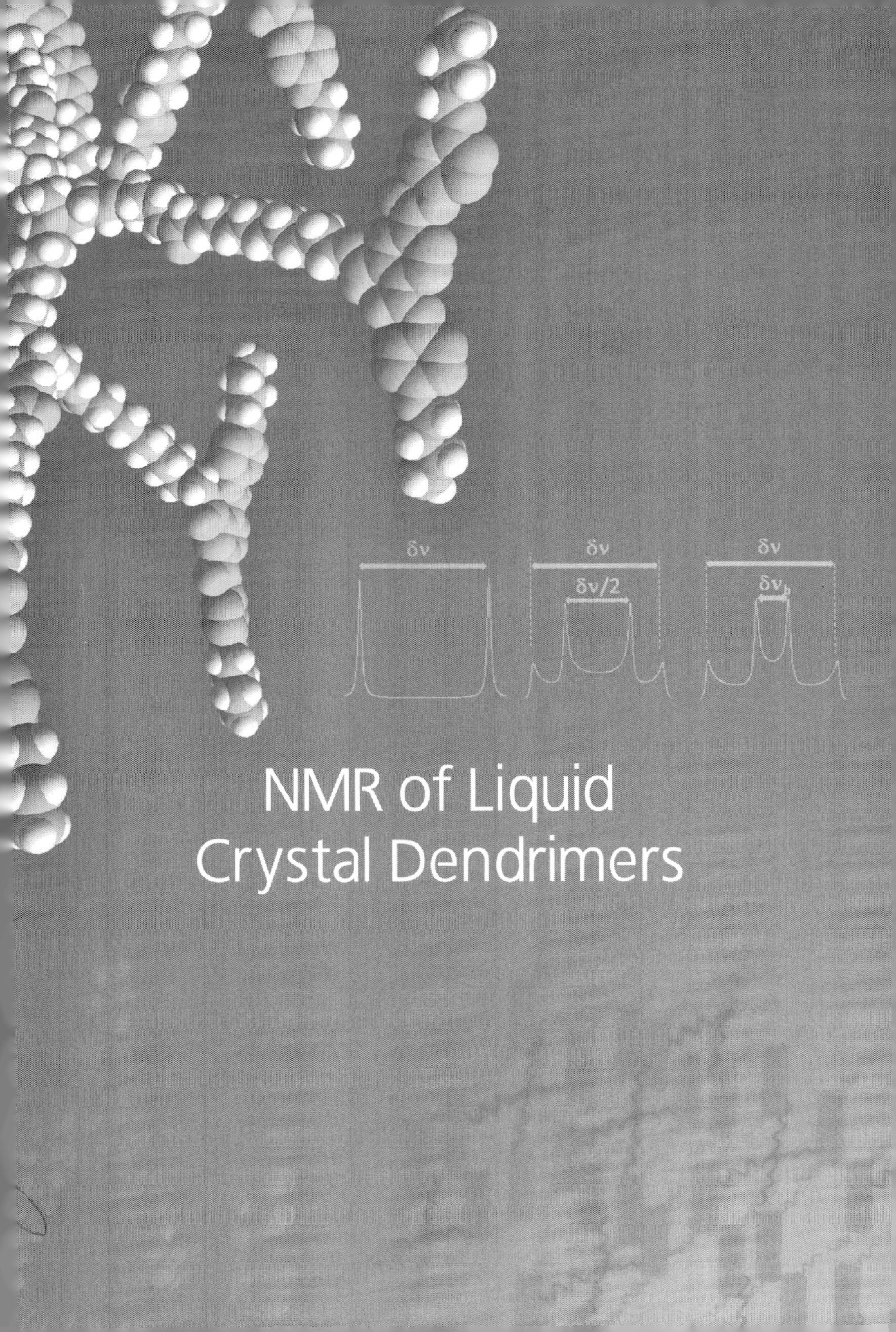
δν
δν
δν/2
δν
NMR of Liquid
Crystal Dendrimers

NMR of Liquid Crystal Dendrimers

Carlos R. Cruz
João L. Figueirinhas
Pedro J. Sebastião

Published by

Pan Stanford Publishing Pte. Ltd.
Penthouse Level, Suntec Tower 3
8 Temasek Boulevard
Singapore 038988

Email: editorial@panstanford.com
Web: www.panstanford.com

British Library Cataloguing-in-Publication Data
A catalogue record for this book is available from the British Library.

NMR of Liquid Crystal Dendrimers

ISBN 978-981-4745-72-7 (Hardcover)
ISBN 978-981-4745-73-4 (eBook)

Printed in the USA

Contents

Preface

This book, about nuclear magnetic resonance (NMR) studies on liquid crystal dendrimers, is expected to be useful for researchers and graduate students in the fields of dendrimers, liquid crystals, and NMR. Besides a general presentation of the structural properties of liquid crystal dendrimers' mesophases, the theoretical principles and experimental techniques of NMR, typically used for the study of these systems, are explained in an accessible way to graduate students of physics, chemistry, and materials science. Applications of these techniques in specific experimental studies are also presented.

Most of the experimental work on NMR spectroscopy and relaxation of liquid crystal dendrimers described herein was done in the framework of collaborations between the authors and several colleagues. The investigations carried out by these chemists and physicists were essential to the accomplishment of the results that were published in the literature and cited, particularly in Chapters 8 and 9. These achievements were possible, in a great part, due to the participation in two *Marie Curie* European projects funded by the European Commission, *LCDD, Supermolecular Liquid Crystal Dendrimers* (HPRN-CT-2000-00016) and *Dendreamers: Functional Liquid Crystalline Dendrimers, Synthesis of New Materials, Resource for New Applications* (FP7-PEOPLE-2007-1-1-ITN 215884), and the project *From Molecular Order and Dynamics to Biaxial Nematic Phases* (PTDC/FIS/65037/2006), funded by the Portuguese Science and Technology Foundation, FCT. The authors are grateful to the institutions that provided financial support to these projects and, above all, to those who took part in the investigations—all the *Marie Curie* fellows, young researchers, and the respective team leaders. Some of these colleagues are co-authors of the cited works, others are members of the teams participating in the projects referred, and all of them, in one way or another, were of key importance due to very enlightening scientific discussions in which the authors had the privilege to participate. For all that, we are especially grateful to Alina Aluculesei, Mariana Cardoso, José Cascais, Bertrand Donnio, Ronald Dong, Leonas Dumitrascu, Gabriel Feio, Daniela Filip, Helena Godinho, Luís Golçalves, Daniel Guillon, Panagiota Karahaliou, Paul Kouwer, Matthias Lehmann, Georg Mehl, Mercedes Marcos, Katarzyna Merkel, Thomas Meyer, Demetri Photinos, Srinath Polineni, Ant´onio C. Ribeiro, Ant´onio Roque, Jean-Michel Rueff, Duarte Sousa, Jose Luis Serrano, Carsten Tschierske, Fabian Vaca Chávez, Alexandros Vanakaras, Alexandra Van-Quynh, Jagdish Vij, Marija Vilfan, and Daniela Wilson. Carlos Cruz also thanks Instituto Superior Técnico of the University of Lisbon for the sabbatical leave that allowed him to dedicate a great part of his time to this project.

We are also deeply grateful to our families, who gently and patiently supported us many times during our absence in the course of the enthusiastic research work.

Carlos R. Cruz
João L. Figueirinhas
Pedro J. Sebastião

Chapter 1

Introduction

1.1 Dendrimers

Dendrimers are molecules with very special properties due to their symmetric hyperbranched structure based on the connection of self-similar chemical groups. The term "dendrimer" is derived from the Greek words *dendros*, which means "tree," and *meros*, meaning "part" (Crampton and Simanek, 2007). In fact, the structure of a dendrimer resembles that of a tree crown. Somehow, dendrimers may be considered as a peculiar variant of polymers where the monomers systematically branch off, generating a tree-like structure, instead of simply connecting linearly to each other forming a single, long flexible-chain characteristic of ordinary polymers. However, contrary to common synthetic polymers, which have significant levels of polydispersity (heterogeneity of molecular weights), dendrimers are, in general, practically monodisperse (i.e., a sample is composed of similar molecules). This important property is related to the particular sequential synthetic procedures used in the preparation of dendrimers.

The dimensions of these molecules, typically of the order of the nanometers, may be synthetically tuned by controlling the number of sequential "layers" of the dendrimer. This number of

NMR of Liquid Crystal Dendrimers
Carlos R. Cruz, João L. Figueirinhas, and Pedro J. Sebastião

ISBN 978-981-4745-72-7 (Hardcover), 978-981-4745-73-4 (eBook)
www.panstanford.com

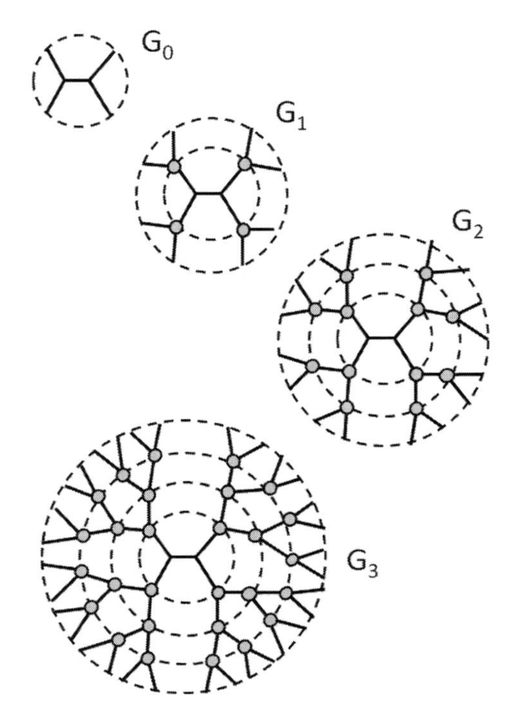

Figure 1.1 Schematic representation of dendrimer structures of generations 0, 1, 2, and 3.

layers, resulting from the levels of branching, from the center to the periphery, defines the generation of the dendrimer (Newkome et al., 2001). Figure 1.1 shows a schematic representation of dendrimer structures of different generations.

Two general approaches described as divergent and convergent may be used in the preparation of those systems. In the divergent method, which was independently introduced by Vögtle and Tomalia in 1978–1979 (Ward and Baker Jr., 2008), the dendrimer molecule is synthesized starting from an inner dendritic core where a layer of branched monomers is attached, followed by a sequence of self-similar steps in which new layers of monomers are successively added until the desired generation is reached (see Fig. 1.2A). In the convergent method proposed by Frechet for the synthesis of

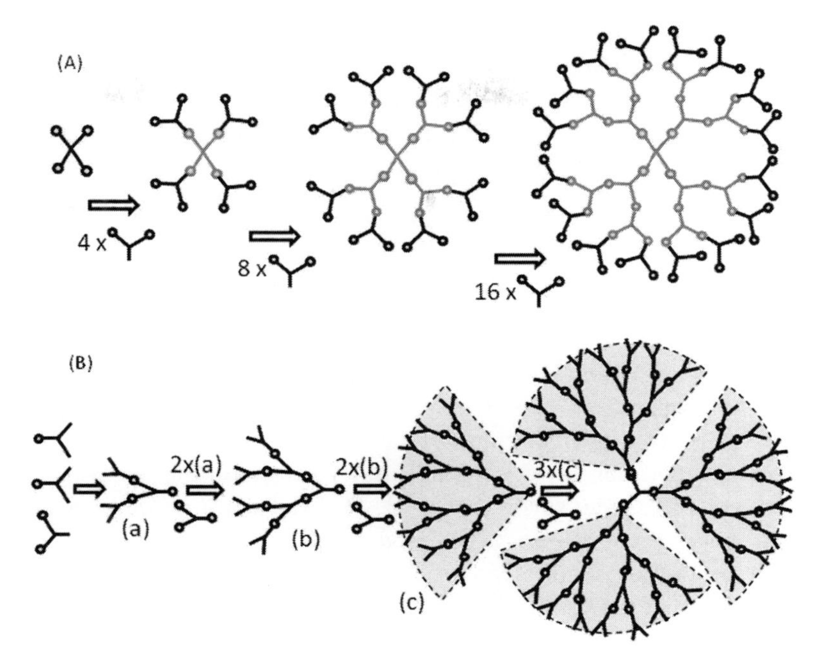

Figure 1.2 Schematic representation of the divergent (A) and convergent (B) dendrimer synthetic approaches.

polyether dendrimers (Hawker and Frechet, 1990), the synthesis follows the scheme presented in Fig. 1.2B, starting from the most outward groups, which are joined to an inner similar segment, forming a more complex branch that is next linked to a similar one by another inner group, following a sequence until a number of complex branches (*dendrons*) are finally joined together to the central core.

Typically, a dendrimer may be divided into a central section called a core, an inner shell comprising the hyperbranched structure of self-similar inner layers, and an outer shell of terminal groups. This structure may be characterized by the following parameters: the multiplicity of the core, N_{C}, the degree of branching, N_{B}, and the generation, G. The number of terminal groups, Z, is given by (Newkome et al., 2001) (see Fig. 1.1).

$$Z = N_{\mathrm{C}} N_{\mathrm{B}}^{G} \tag{1.1}$$

The total number of junctions in a dendrimer may be calculated by adding the links starting from the core ($k = 0$) to the generation $G - 1$ ($k = G - 1$) (notice that the upper generation, $k = G$, corresponds to the outer surface of the dendrimer where the links are open and therefore not corresponding to junctions):

$$\begin{aligned} J &= N_C + N_C N_B + N_C N_B^2 + \ldots + N_C N_B^{(G-1)} \\ &= N_C \sum_{k=0}^{(G-1)} N_B^k = N_C \left[\frac{(N_B^G - 1)}{N_B - 1} \right] \end{aligned} \tag{1.2}$$

The fact that the number of terminal groups and the degree of arborescence defined by number of junctions J grow exponentially with the generation G sets a limit to the perfect starburst dendrimer growth (without defects) due to a congestions effect. At a certain critical generation G_C, the available outer surface of the dendrimer (solvent-accessible surface area A_{SAS}) (Newkome et al., 2001) will not be enough to accommodate all the branches of the next generation. This means that further growing will force the emergence of defects in the dendrimer structure. The external area of the dendrimer A_{SAS} follows the proportional relationship

$$A_{SAS} \approx d^D \tag{1.3}$$

where d represents the linear dimensions of the dendrimer (roughly proportional to G) and D is the fractal dimension of the dendrimer surface, which is equal to 2 for a perfect sphere and between 2 and 3 for more complex objects (e.g., for sixth-generation polyamidoamine (PAMAM) $D = 2.42$) (Newkome et al., 2001). For low-generation dendrimers, $G << G_C$, the available external surface is much larger than the area effectively occupied by the dendrimer terminal groups and therefore dendritic structures tend to be eventually flexible and of variable shape, depending on the steric properties of the chemical groups involved. Higher-generation dendrimers tend to progressively assume a globular shape, especially for generations close to G_C.

Important technological applications of these systems, both in biomedicine and in materials science, have been proposed as, for instance, gene transfer and drug delivery agents, nanocatalysts, and contrast agents for nuclear magnetic resonance (NMR) imaging. Most of these applications are based on the functionalization of the outer shell of the dendrimer with chemically adequate terminal segments (Astruc et al., 2010).

1.2 NMR of Liquid Crystals and Liquid Crystal Dendrimers

Liquid crystals are materials that combine the ability to flow (and therefore adapt their shape to the confining surfaces) with the anisotropic physical properties typical of crystalline substances. The molecules of these materials are less organized than the conventional crystals and more ordered than common isotropic liquids. This dual behavior, combined with their anisotropic interaction properties with electromagnetic fields and surfaces, is the basis of most of the applications of liquid crystals, such as the ubiquitous liquid-crystal displays (LCDs).

Liquid-crystalline dendrimers are fascinating materials that join the characteristics of dendrimers with the anisotropic physical behavior and molecular self-organization typical of liquid crystals. This unique association of physical and chemical properties, together with the possibility of multiselective functionalization put forward by dendrimers, opens new technological perspectives. Ionic conductivity, dendritic photochromic-LC actuators, light valves, flexible LCDs and memory devices based on magnetic dendrimers, sensors based on receptor units, and smart catalysts to be switched on/off by light are examples of such potential for applications.

NMR is a powerful experimental technique applied in materials science and an important tool to the study of molecular organization and dynamics. This spectroscopic technique results from the interaction of radio-frequency (RF) radiation with specific nuclear spins present in the materials (e.g., ^{1}H, ^{2}H, ^{13}C, and many others).

As will be detailed in forthcoming chapters, several interactions contribute as perturbations to the nuclear Zeeman effect, which occurs when a sample is placed in the presence of an external static magnetic field. Some of these interactions are anisotropic and therefore particularly useful in the study of materials where molecular order plays in important role, such as liquid crystals.

This book presents an introduction to dendrimers properties with special insight on liquid crystal dendrimers and a detailed description of the NMR theory and experimental techniques used in the investigation of these materials. Representative NMR research experimental results on liquid crystal dendrimers are reviewed, providing a view of the application of NMR techniques to the investigation of these interesting systems.

In Chapter 2, general fundamental concepts of liquid crystals will be presented, with special emphasis on the molecular order and the structural properties of the different phases exhibited by these materials.

In Chapter 3, the molecular architecture and phase structures of liquid crystal dendrimers will be introduced, having in mind some general properties such as molecular flexibility and shape and their influence on the liquid-crystalline phase organization and molecular ordering and dynamics, which are accessible by NMR techniques. Specific physical properties (like viscoelastic constants, dielectric permittivities, magnetic susceptibilities, and refractive indices), which are particularly important for applications, depend on the phase organization and molecular motions, justifying the relevance of NMR for investigation of these materials.

In Chapter 4, the physical principles of NMR will be introduced. Fundamental concepts will be presented and put into perspective as part of the scientific development of the second half of the twentieth century.

The theory and experimental aspects of NMR spectroscopy of anisotropic physical systems will be discussed in Chapter 5. Both the theory introduced in Chapters 4 and 5 and the experimental techniques presented in Chapter 5 will be particularly useful for the study of molecular ordering of liquid-crystalline dendrimers.

NMR relaxometry is an experimental technique based on the observation of the process of reestablishment of thermal equilibrium of a spin system previously perturbed by RF electromagnetic radiation. This process, based on the energy transfer between the perturbed spin system and the thermal degrees of freedom (usually called lattice), is strongly influenced by the molecular motions. Consequently, molecular dynamics becomes accessible through this type of NMR studies. Chapter 6 of this book presents a theoretical introduction to NMR relaxometry and molecular dynamics of spin systems.

In Chapter 7, different experimental aspects of NMR relaxometry are presented. Both Chapters 6 and 7 are especially useful to the development of subsequent chapters on molecular dynamics of liquid-crystalline dendrimers relaxometry studies.

Finally, in Chapters 8 and 9, several experimental case studies of NMR spectroscopy and relaxometry of liquid-crystalline dendrimers are, respectively, reviewed. Those selected examples are representative of the authors' contributions to the field and intend to provide to the reader, as much as possible, a practical insight into the application of the theory and experimental techniques introduced in the previous chapters.

Chapter 2

Liquid Crystals

2.1 Liquid Crystals

In liquid crystal phases, molecules are organized with an intermediate degree of order, between the 3D positional long-range order exhibited by crystals and the short-range ordering typical of liquids (a more precise definition of the liquid crystal types of ordering will be given ahead). These intermediate phases are also called mesophases. The molecules or molecular segments, which give rise to this type of order are called mesogenic or mesomorphic.

As equilibrium thermodynamic phases, mesophases formed by liquid-crystalline (LC) molecules exist within determined pressure and temperature limits. A material that exhibits several of those phases is called polymorphic and this behavior is referred as polymorphism.

Liquid crystals may be classified into two major categories: thermotropics and lyotropics. In thermotropic liquid crystals, the transition between different mesophases (for a given pressure, usually the atmospheric pressure) is determined by temperature changes. Lyotropic liquid crystals are solutions of specific molecules (usually surfactants) in an appropriate solvent (most commonly

NMR of Liquid Crystal Dendrimers
Carlos R. Cruz, João L. Figueirinhas, and Pedro J. Sebastião

ISBN 978-981-4745-72-7 (Hardcover), 978-981-4745-73-4 (eBook)
www.panstanford.com

water) and the phase transitions (at given temperature and pressure) are caused by variation of the concentration.

Typically, when the solvent is water, lyotropic compounds are amphiphilic, that is, they result from the junction of a hydrophilic (polar) and a hydrophobic (aliphatic) segment in the same molecule. This type of LC phases are formed due to the energy–entropy balance resulting from the organization of the surfactant molecules in order to avoid the contact of the hydrophobic part of the compound with the water molecules.

At the lowest concentration for which this effect occurs (critical micellar concentration) the mesogenic molecules form micelles (generally of spherical shape) with the polar part at the outside, in contact with water, and the aliphatic part inside. As the concentration increases, other progressively more complex phases may occur, with the surfactant molecules organized in layers (lamellar phases) in columns (columnar phases), in cubic and eventually other three-dimensionally organized phases. In Fig. 2.1 a schematic representation of some lyotropic phases is presented.

Historically, lyotropic liquid crystals were the first to be discovered. Actually, the type of molecular organization characteristic of this type of materials is omnipresent in living matter and their investigation begun in the middle of the nineteenth century by R. Virchow (1854) who was studying the properties of myelin. This material is formed from lipids and proteins of biological origin and is present in particular in the nervous system (Petrov, 1999).

Some lyotropic compounds also exhibit thermotropic LC phases when in pure state (without solvent). These materials are called

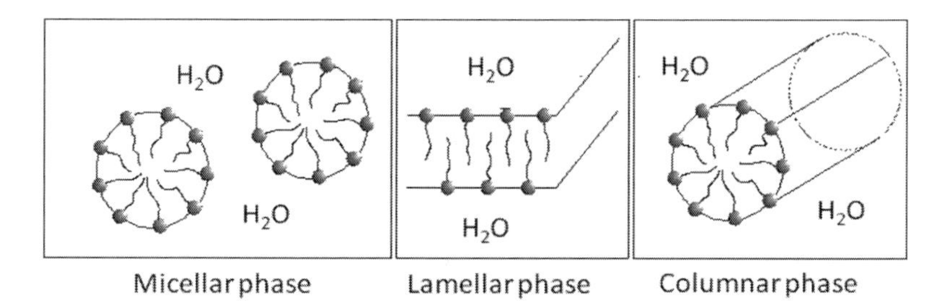

Figure 2.1 Schematic representation of micellar, lamellar, and columnar lyotropic LC phases.

amphitropic. Another important class of liquid crystals, which in a broad sense may be considered as a special case of lyotropics, is the so-called chromonics. In these systems, molecules, which are usually of flat shape, pile up into stacks of different length that constitute the mesogenic units, which, in their turn, organize into more complex phases within a solvent (e.g., water) (Lydon, 2010).

Thermotropic liquid crystals are synthetic compounds, nowadays with a huge industrial impact, greatly due to their generalized use in the display technology but also in other areas such as temperature and opto-electronic sensors (Blinov and Chigrinov, 1994) and biomedicine (Crawford and Woltman, 2007).

The first example of this type of compounds was discovered in 1888 by a botanist (F. Reinitzer) in a cholesterol derivative (cholesterol benzoate) (Gray et al., 2008; Reinitzer, 1888; Sluckin et al., 2004). Reinitzer detected an unknown phase (between temperatures corresponding to what he called two fusion points) with the appearance of a translucent liquid. That phase (mesophase) was later identified by the physicist Otto Lehmann, using microscopy with polarized light, as a material with mechanical properties of a liquid (fluidity, incapability of sustaining stress) and optical characteristics of a crystal (anisotropic refractive index) (Gray et al., 2008; Lehmann, 1889; Sluckin et al., 2004). The term "liquid crystal" ("Flüssige Krystalle" in German) was first introduced by Lehmann.

Both thermotropic and lyotropic liquid crystals may be composed of low-molecular-weight molecules, which coincide with the mesogenic units of the system, or more complex systems like polymers, oligomers or dendrimers. In the later, the mesogenic units are structural elements of a macromolecule (polymer liquid crystal) or specific parts of an oligomer or dendrimer (Donnio et al., 2007). In Fig. 2.2, schematic representations of low-molecular-weight and polymer liquid crystals are presented. Polymer liquid crystals may be classified as main-chain or side-chain crystals when the mesogenic units are included in the polymeric chain or laterally disposed with respect to it, respectively, as shown in Fig. 2.2 This book concerns thermotropic LC dendrimers (also schematically depicted in Fig. 2.2), whose structure will be discussed further on.

The molecules of thermotropic low-molecular-weight liquid crystals (or the mesogenic units, in the case of more complex

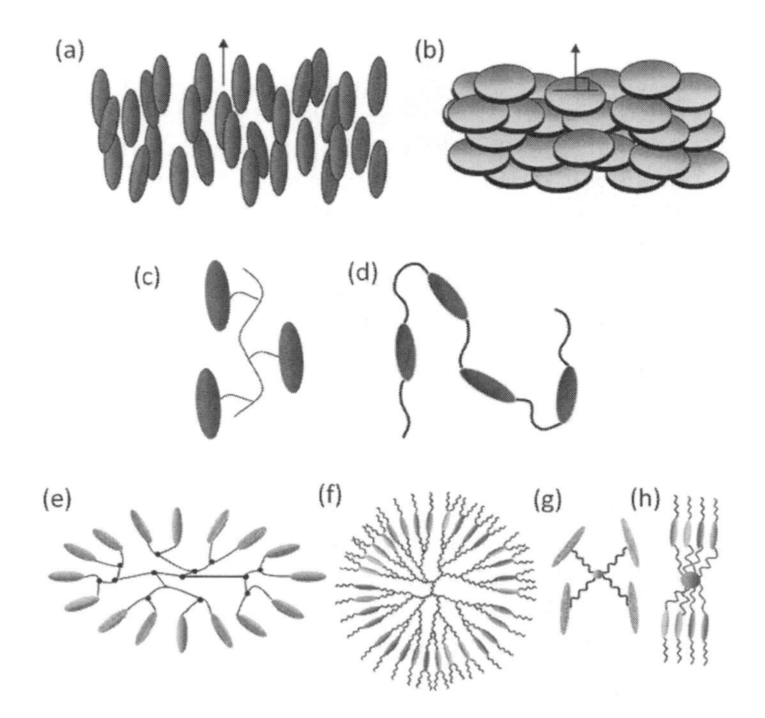

Figure 2.2 Schematic representation of low-molecular-weight polymer and dendrimer liquid crystal molecules. (a) Calamitic nematic, (b) discotic nematic, (c) side-chain LC polymer, (d) main-chain LC polymer, (e) G2 LC dendrimer, (f) G3 LC dendrimer, (g) G0 LC dendrimer (tetrapode), and (h) G0 LC dendrimer (octopode).

systems like polymer or dendrimer liquid crystals) are in general "anisometric" objects, that is, they have one of the dimensions significantly larger or smaller than the others. These molecular entities (mesogens) may have different shapes and, as will be discussed ahead, this will strongly influence the symmetry (the spatial molecular arrangement) of the phases exhibited by the material. Most commonly those mesogens have rod-like (calamitic) or disc-like (discotic) shapes but other examples are also important like lath-like (sanidic) (Haristoy et al., 2000), polycatenar (elongated with multiple terminal chains) (Malthete et al., 1985, 1986; Tinh et al., 1986), bent-shape (Pelzl et al., 1999), pyramidic (Malthete and Collet, 1985; Zimmermann et al., 1985), bowlic (Leung and

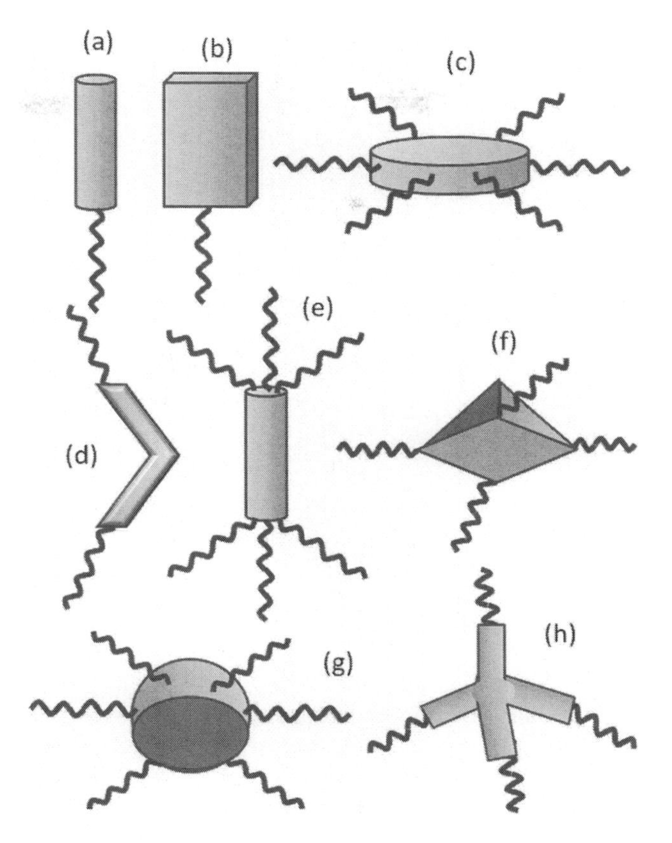

Figure 2.3 Schematic representation of LC mesogenic elements ideal shapes: (a) calamitic, (b) sanidic, (c) discotic, (d) bent shape, (e) polycatenar, (f) pyramidic, (g) bowlic, and (h) tetrahedric.

Lin, 1987; Lin, 1987), macrocyclic (Bonsignore et al., 1990), and tetrahedric (Eidenschink et al., 1990), as shown in Fig. 2.3.

The LC phase more similar to ordinary (isotropic) liquids is the nematic phase. In this phase the molecules tend to align an axis (in the case of calamitic molecules this axis corresponds approximately to the direction of longest molecular dimension) along a common direction, keeping their geometric centers randomly distributed in space as in an isotropic liquid (see Fig. 2.4). This type of order is called orientational order and its existence is a property of all LC

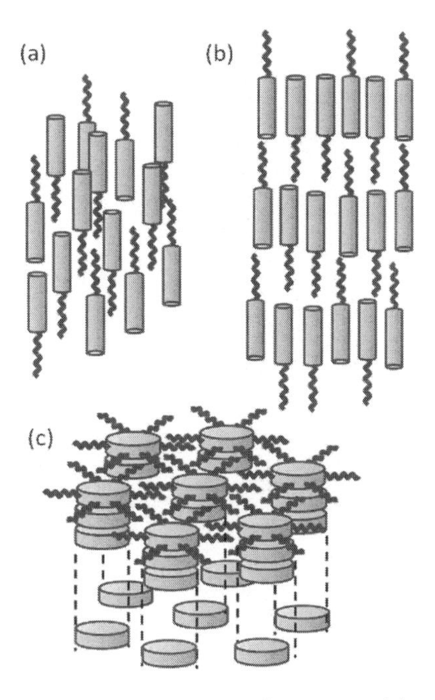

Figure 2.4 Schematic representation of nematic (a), smectic (b), and columnar (c) phases of thermotropic liquid crystals. In the case of smectic (lamellar) and columnar mesophases microsegregation between rigid and flexible molecular segments is shown.

phases (both lyotropic and thermotropic). In most cases, this phase follows the isotropic liquid phase when the temperature is lowered. The transition between the highest temperature mesophase and the isotropic phase is called the clarification point because the mesophase (most commonly nematic) that has a blurry appearance turns into a transparent liquid when this temperature is reached on heating.

Another important property of most of the thermotropic LC molecules is the presence of elements of different chemical nature, usually a rigid segment (typically of aromatic type) and a flexible part, most commonly, an aliphatic chain. Other types of segments may be included, for example, polar groups (CN, NH_2, NO_2) or siloxane groups. The effect of microsegregation between these different types of elements plays an important role on the molecular

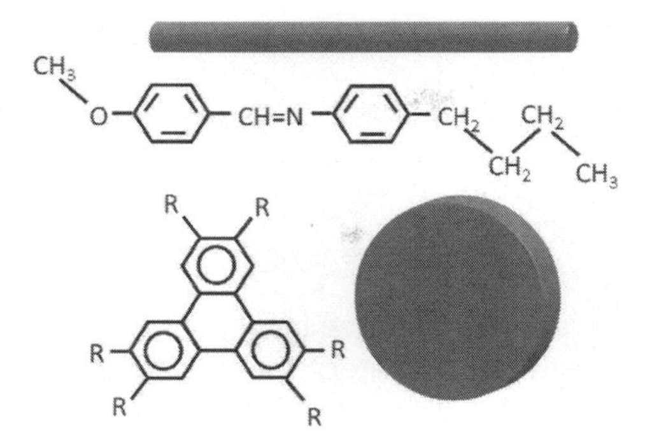

Figure 2.5 Examples of typical calamitic and discotic compounds. In the case of the discotic, R is typically a flexible aliphatic chain, C_nH_{2n+1}.

organization of LC phases (Skoulios and Guillon, 1988; Tschierske, 1998). Below a given transition temperature, molecular segments of similar chemical nature tend to be close to each other in the mesophase leading, for instance, to the formation of layers (lamellar, or smectic phases), in the case of calamitic molecules or columns (columnar phases), in the case of discotic molecules as schematically shown in Fig. 2.4. Nematic, smectic and columnar phases exist in low-molecular-weight, polymeric, and dendritic liquid crystals. In certain special cases, LC phases exhibit more complex 3D positional order types, like cubic, orthorhombic and others (Goodby et al., 2008).

Examples of typical calamitic and discotic compounds giving rise respectively (a) to nematic or smectic and (b) columnar phases are presented in Fig. 2.5.

2.2 Mesophases of Thermotropic Liquid Crystals

2.2.1 Nematic Phases

The nematic phase, referred above, may appear both in thermotropic and lyotropic liquid crystals, and it may be exhibited by low-molecular-weight polymer and dendrimer LC materials. It is also important to notice that molecules of different shapes (calamitic, discotic, sanidic, bent core, ...) may originate nematic phases.

The long-range orientational order characteristic of nematics (and also present in all LC phases) may be defined by the ordering matrix (de Gennes and Prost, 1993):

$$S_{\alpha\beta}^{ij} = \left\langle \frac{1}{2}\left(3\cos\theta_{i\alpha}\cos\theta_{j\beta} - \delta_{\alpha\beta}\delta_{ij}\right)\right\rangle \tag{2.1}$$

where the brackets stand for ensemble average and $\theta_{i\alpha}, \theta_{j\beta}$ represent the angles between the laboratory frame axes i,j = (X, Y, Z) and a fixed molecular frame axes α, β = (x, y, z), as shown in Fig. 2.6. In the most general case the ordering matrix has 34 independent parameters (by definition it is a symmetric and traceless tensor) (Straley, 1974).

The simplest case corresponds to the uniaxial nematic phase of cylindrically symmetric molecules, where the molecular organization can be defined by a single axis in space, coinciding with the average preferential orientation of the molecules. In that case, the elements of the ordering matrix in Eq. 2.1 is reduced to a single significant parameter $S = S_{zz}^{ZZ}$ called the nematic order parameter, originally introduced by Tsvetkov (Tsvetkov, 1942):

$$S = \left\langle \frac{1}{2}\left(3\cos^2\theta - 1\right)\right\rangle \tag{2.2}$$

In Eq. 2.2 θ represents the angle between the molecular principal axis of a molecule (with average cylindrical symmetry) and a single direction of preferential molecular alignment, defined by a unit vector **n** called director (see Fig. 2.6).

It might happen that a single axis is not sufficient to define the orientational order of the phase. That situation corresponds to an important variant of the nematic phase, the biaxial nematic, N_b, theoretically predicted by Freiser in 1970 (Freiser, 1970). In that case, besides the orientational order defined by the main director **n**, the molecules are oriented in average along an additional preferential direction in the plane perpendicular to **n** (see Fig. 2.7). The biaxial nematic ordering requires the introduction of additional orientational order parameters. Following the procedure introduced by Straley (Straley, 1974) four parameters may be defined on the basis of the ordering matrix elements (see Fig. 2.6 and Eq. 2.1).

$$S = S_{zz}^{ZZ} \tag{2.3}$$

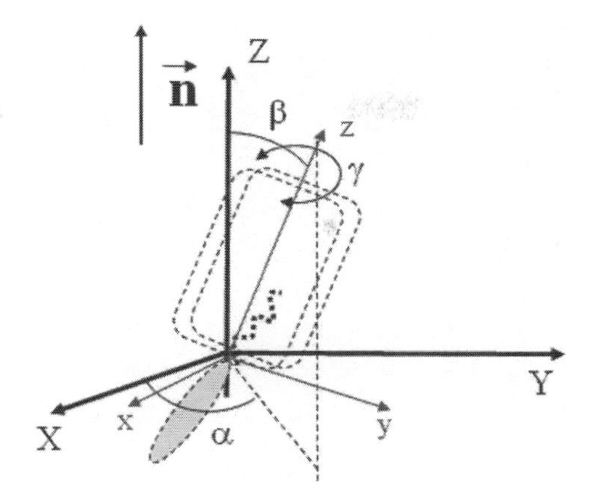

Figure 2.6 Laboratory and molecular frames used in the definition of the ordering matrix.

$$D = S_{xx}^{ZZ} - S_{yy}^{ZZ} \tag{2.4}$$

$$P = S_{zz}^{XX} - S_{zz}^{YY} \tag{2.5}$$

$$C = (S_{xx}^{XX} - S_{yy}^{XX}) - (S_{xx}^{YY} - S_{yy}^{YY}) \tag{2.6}$$

The meaning of those order parameters is the following:

S is the main nematic order parameter, previously defined, expressing the tendency of alignment of the main molecular axis with respect to the main director **n**.

D reflects the "molecular biaxiality" resulting from the molecular anisotropy in the plane perpendicular to the main molecular axis z. It is important to notice that a value of $D \neq 0$, associated with some degree of molecular "flatness" (see Fig. 2.7), doesn't mean that the phase is biaxial, but just that the molecular orientational distribution is not isotropic in the xy plane of the molecular frame. If the distribution of the orientations of those molecules is symmetric with respect to rotations in the XY laboratory frame, then the phase is still uniaxial.

P and C order parameters express the phase biaxiality resulting from a possible tendency of the molecules to orient anisotropically in the plane XY of the laboratory frame. If the phase is biaxial, additional secondary directors, **l** and **m**, are defined in the plane

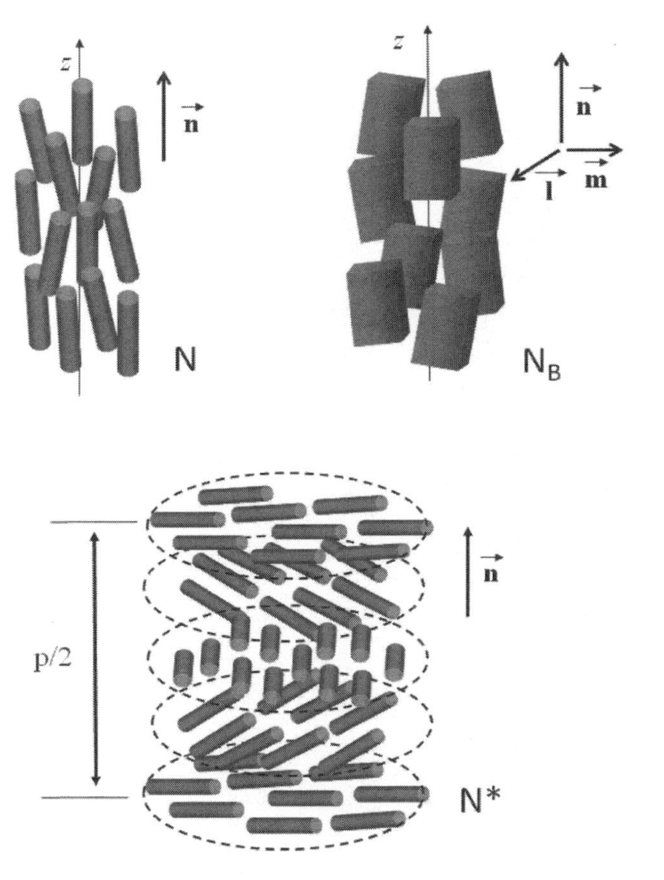

Figure 2.7 Schematic representations of nematic (N), cholesteric (N*), and biaxial nematic (N_b) phases.

perpendicular to **n** and the biaxial orientational order results from the tendency of the molecules to assume some preferential average orientation with respect to those secondary directors (additionally to the main nematic order defined by **n**).

Considering now, for simplicity, the case of the uniaxial nematic phase where the molecules tend to align, in average, along a single direction, any distortion of the molecular alignment from this spontaneous orientation, costs a certain amount of energy. Deviations of the director (or director field, considering that **n** varies

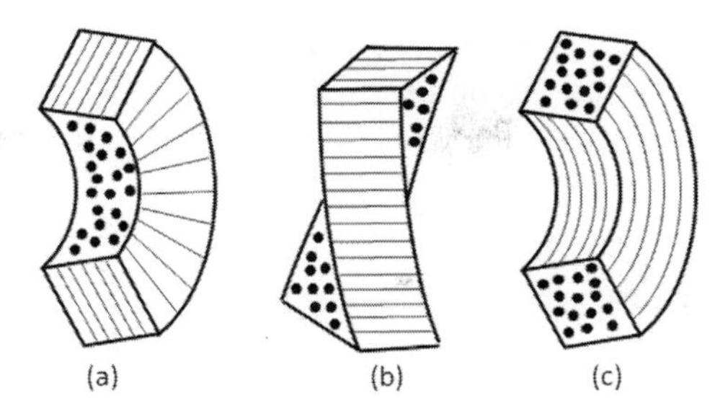

Figure 2.8 Elementary modes of nematic elastic distortions: splay (a), twist (b), and bend (c).

continuously as a function of space coordinates) may be caused by impurities or interaction with surfaces or external fields imposing certain preferential molecular orientation.

The associated mechanism of "curvature elasticity" may be described by the so-called Frank free-energy density in terms of three basic distortion modes: splay, twist, and bend, characterized by three elastic constants, K_1, K_2, and K_3, respectively. The corresponding expression is:

$$F = \frac{1}{2}K_1\,[\nabla.\mathbf{n}]^2 + \frac{1}{2}K_2\,[\mathbf{n}.\,(\nabla \times \mathbf{n})]^2 + \frac{1}{2}K_3\,[\mathbf{n} \times (\nabla \times \mathbf{n})]^2 \quad (2.7)$$

Additional terms are added to this expression when the nematic material interacts with electromagnetic or surface force fields. In that case, the spatial configuration of the director field is the result of the competition between the elastic curvature energy and the external interactions, and is obtained by the minimization of the free energy, including the referred additional terms (de Gennes and Prost, 1993).

A schematic representation of the elementary modes of nematic distortion is represented in Fig. 2.8.

Some LC compounds with an element of chirality in their molecules give rise to a variation of the nematic phase called "cholesteric" or "chiral nematic" (N*). The N* phase is locally similar to the nematic phase but the local average orientation of the molecules (local director) follows a helix with the axis (cholesteric director) perpendicular to the local director (see Fig. 2.7). The Frank free energy for that phase is given by an expression similar to Eq. 2.7 where the term corresponding to the contribution of the twist distortion is now given by $1/2\ K_2[\mathbf{n}.(\nabla \times \mathbf{n}) + q]^2$. The parameter $q = 2\pi/p$ expresses the natural pitch of the phase ensuring that F equals zero for the spontaneously twisted configuration of the cholesteric (de Gennes and Prost, 1993).

2.2.2 Smectic Phases

In smectic liquid crystals, the molecules are organized in layers, meaning that, besides the long-range orientational order previously described, the system presents long-range positional ordering in one dimension (corresponding to the direction perpendicular to the layers).[a]

In uniaxial smectics of disordered layers (in most cases formed by rod-like mesogens), such as smectic A (SmA) where the long molecular axis are in average perpendicular to the layers' plane, the positional order may be described by the density function

$$\rho = \rho_0 + Re\left\{\psi.e^{i\frac{2\pi}{d}z}\right\} \tag{2.8}$$

This expression correspond to the first-order Fourier series approximation of the periodic distribution of density in the direction perpendicular to the layers, z (see Fig. 2.9). ρ_0 is the average density, d is the layers' spacing and the smectic order parameter, $\psi = |\psi|e^{i\varphi}$, is a complex quantity, where $|\psi|$ represents the amplitude of the density variation in the direction perpendicular to the layers and φ determines the layers position (de Gennes and Prost, 1993).

In smectics with disordered layers and tilted molecules, smectic C, an additional order parameter must be included, expressing the molecular tilting and its direction (see Fig. 2.9):

$$\chi = \alpha e^{i\phi} \tag{2.9}$$

where α is the tilt angle with respect to the layers' normal, and ϕ is the azimuthal angle, which defines the orientation of the tilting in

[a]More precisely, the positional order in smectic phases must defined as quasi-long range, since the Landau–Peierls instability (logarithmic increase of thermal fluctuations with distance in 1D systems) forbids the existence of long-range positional order in one dimension. Nevertheless, in practical terms, the number of layers in a smectic domain is high enough to ensure its macroscopic size (Landau et al., 1986).

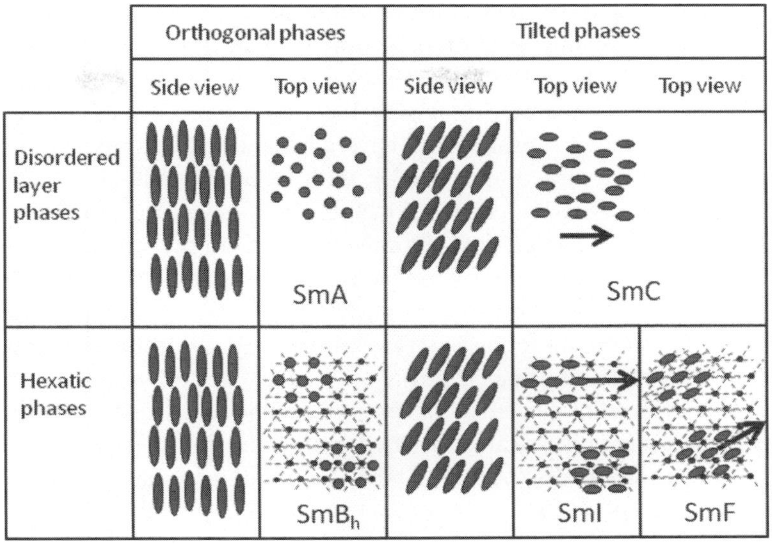

	Orthogonal phases		Tilted phases		
	Side view	Top view	Side view	Top view	Top view
Disordered layer phases		SmA		SmC	
Hexatic phases		SmB_h		SmI	SmF

Figure 2.9 Schematic representation of different types of smectic liquid crystal phases.

the layer's plane. Here, a director (the $\vec{c}$ director) may be defined in the layer's plane, which becomes analogous to a 2D nematic system.

In both SmA and SmC phases, the centers of mass of the molecules are randomly distributed within the layers, forming what may be described as a 2D liquid.

Due to the constraints imposed by the additional 1D positional ordering, the free energy in smectic phases assumes an expression different from the previously introduced for the nematics. In a first-order approximation this expression is given by:

$$n_x = -\frac{\partial u_z}{\partial x}; n_y = -\frac{\partial u_z}{\partial y}; n_z \simeq 1$$

$$F = \frac{1}{2} B \left(\frac{\partial u_z}{\partial x} \right)^2 + \frac{1}{2} K_1 \left(\frac{\partial^2 u_z}{\partial x^2} + \frac{\partial^2 u_z}{\partial y^2} \right)^2 \qquad (2.10)$$

where $u(r)$ describes the displacement of a point with respect to the unperturbed position, K_1 is the bend elastic constant, and B is the elastic constant related to the compression of the layers (Landau et al., 1986).

Some smectic phases present a special type of order in the layers plane (bond orientational order) characterized by short-range positional order (in a hexagonal lattice) and long-range orientational order of the axis of the local lattice bonds. Those phases are called "hexatic" and have different variants depending on the relative direction of the tilt with respect to the hexagonal lattice (SmB_H, SmF, SmI, and SmL). In the SmB_H phase the molecular long axis is perpendicular to the smectic layers. In the SmI and SmF phases the molecules are tilted toward the hexagons vertex and the bisector of the hexagons side, respectively (see Fig. 2.9). In the SmL phase the tilting direction is intermediate of those of the SmF and SmI phases (Goodby et al., 2008).

Some lamellar phases with long-range positional order within the layers were historically classified as smectics. In a more rigorous way, those phases (B, G, J, E, H, K) may be classified as "soft crystals" as, in spite of the long-range positional order inside the layers, they differ from conventional crystals by a considerable degree of molecular disorder, allowing some local reorientation of molecules in the lattice and the possibility of sliding of the layers with respect to each other (Goodby et al., 2008).

Smectic phases of mesogens with strong terminal electric dipoles (e.g., a CN group) may exhibit different peculiar layer spacing arrangements due to competition between smectic order (see Eq. 2.8) and order associated with dipole–dipole coupling, characterized by an antiferroelectric order parameter. Those phases (SmA_1, SmA_2, SmA_d, $Sm\tilde{A}$, SmC_1, SmC_2, SmC_d, $Sm\tilde{C}$) are also called "frustrated smectic" (see Fig. 2.10). Due to its special characteristics, those systems give rise to the reentrance phenomenon characterized by the reappearance of nematic phases (reentrant nematics) for temperatures below those of the smectic phases. The molecular organization in a variety of SmA phases with strong terminal dipoles is schematically shown in Fig. 2.10 (Prost, 1984).

Besides the N* phase, smectic chiral phases may also appear when a chiral element is introduced in the molecular structure, for example, SmC*, SmI*, SmF*. In these phases, a helical disposition of the tilted molecules with the helical axis perpendicular to the smectic layers is induced by the nature of molecular constraints, where chirality plays a very important role. Due to the structural

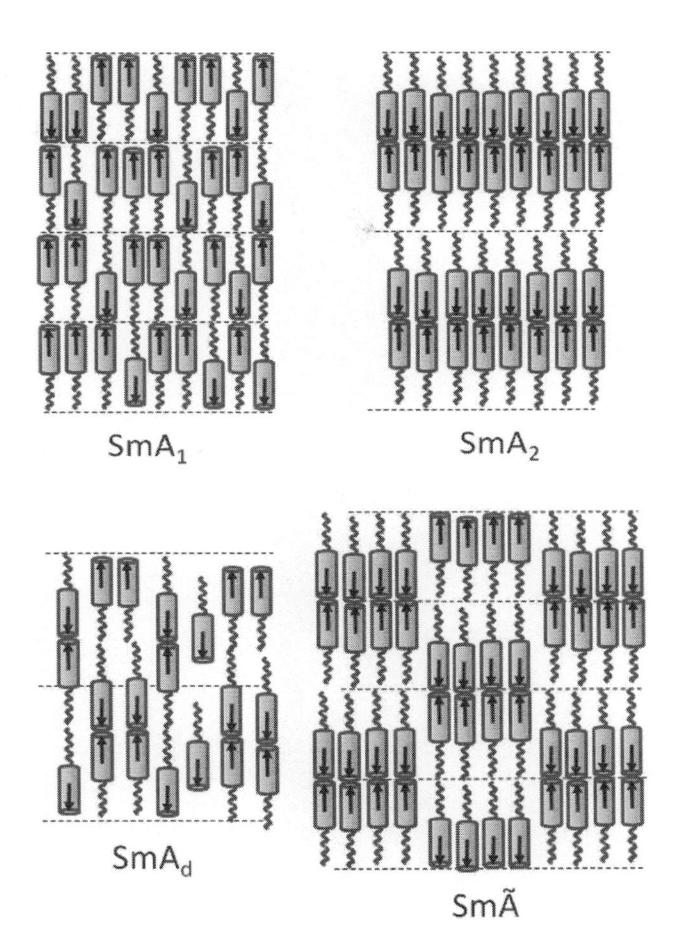

Figure 2.10 Schematic representation of different types of peculiar smectic A phases exhibited by liquid crystals of molecules with strong terminal dipoles SmA_1 (monolayer), SmA_2 (bilayer), SmA_d (partial bilayer), and $Sm\tilde{A}$ (modulated smectic phase). The corresponding smectic C phases (SmC_1, SmC_2, SmC_d, $Sm\tilde{C}$) exhibit similar arrangements of tilted molecules with respect to the layers' normal.

constraints of such phases, lateral dipoles present in the molecules are aligned within each layer, making it ferroelectric (see Fig. 2.11) (Goodby et al., 2008). Tridimensional ferroelectric behavior may be induced by specific surface confinement conditions that force the

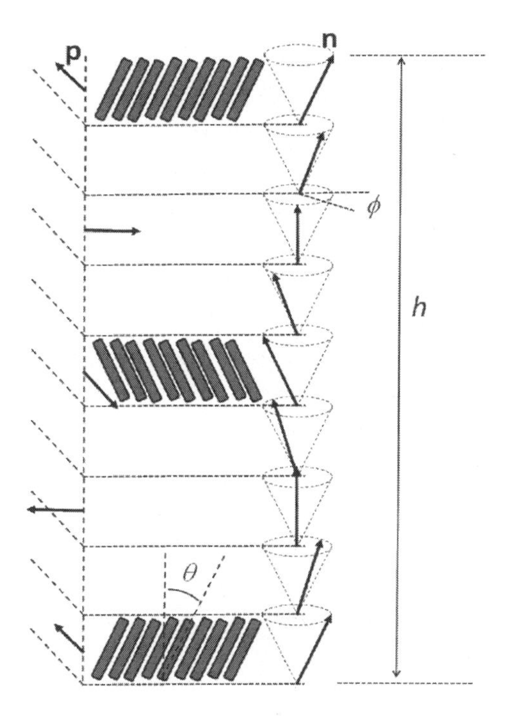

Figure 2.11 Schematic representation of the SmC* phase. **n** is the director, tilted at an angle θ with respect to the normal of the layers and with a tilting direction (ϕ) that rotates from layer to layer, forming a helix of pitch h. The polarization **p** is parallel to the layers, typically perpendicular to the tilting direction in each layer.

helix to unwind. Such an arrangement is known as surface-stabilized ferroelectric liquid crystal (SSFLC) structure and can be used in display applications (Blinov and Chigrinov, 1994).

The inclusion of lateral dipoles is such chiral molecules can also induce the possibility of a variety of complex molecular arrangements like the twisted grain boundary (TGB) phases and blue phases. TBG (TGBA, TGBC) are variants of smectics where blocks of SmA or SmC rotate around a helix with the axis parallel to the layers (see Fig. 2.12).

Blue phases are another type of chiral phases composed of cubic superstructures formed by double-twisted columns of chiral molecules (de Gennes and Prost, 1993).

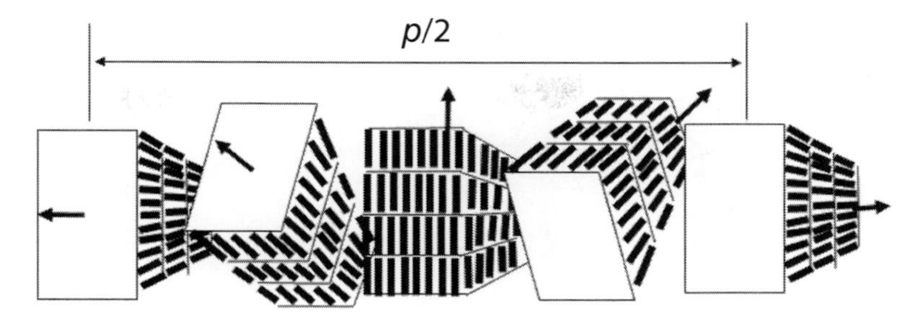

Figure 2.12 Schematic representation of the TGBA phase. The molecules form a local SmA arrangement with a director that rotates along a helical structure of pitch *p* with an axis perpendicular to the local director (parallel to the layers' planes). The TGBC phase is similar to the TGBA phase, with the molecules tilted with respect to the local layers' normal.

The bent-shape molecules ("banana like" or "boomerang like") imply the inclusion of additional elements of molecular orientational order (in comparison with calamitics). In particular the apex of the bent molecular core gives the possibility of defining an orientation in the plane perpendicular to a longitudinal long molecular axis. This special molecular geometry gives rise to rich polymorphisms of the so-called "banana" phases (B1, . . . , B7), which are particular cases of modulated smectics. If lateral dipoles are included, the orientation of the molecules in the layers can be electrically switched giving rise to ferroelectric behavior (Pelzl et al., 1999). Another interesting behavior of those systems is the appearance of biaxial nematic cluster phase (Dingemans et al., 2013).

2.2.3 Columnar Phases

Besides the already mentioned nematic discotic phase, disc-like mesogenic molecules can give rise to columnar phases. In those phases, molecules are arranged in columns, which are organized in 2D lattices in the plane perpendicular to the columnar axis. The long-range positional order in two dimensions is associated with five possible 2D crystalline lattices (Col_h; Col_r, $P2_1/a$, P_2/a, $C2/m$; or oblique Col_{ob}, may appear with ordered or disordered columns) (see Fig. 2.13). Additional rotation and glide reflection symmetry operations led to 17 possible 2D arrangements corresponding to

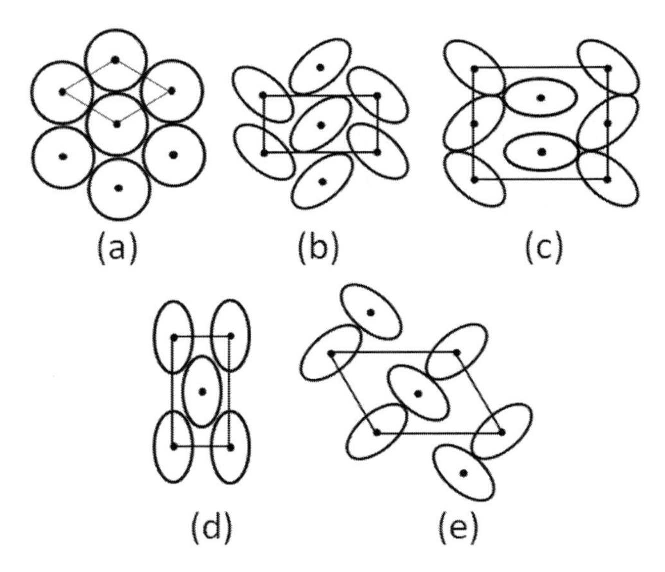

Figure 2.13 Schematic representation of different types of columnar phases. The figures show the transverse section of the columnar structure; the ellipses correspond to the transverse section of a column with tilted molecules (in the case of discotic mesogens): (a) hexagonal Col_h, P6 2/m 2/m; (b) rectangular Col_r, $P2_1/a$; (c) rectangular Col_r, P_2/a; (d) rectangular Col_r C2/m; and (e) oblique Col_{ob}, P_1.

17 2D space groups (Chandrasekhar and Ranganath, 1990; Guillon, 1999; Levelut, 1983).

Transverse sections of the columns may be formed by a discotic molecule, or several molecules (like in polycatenar or biforked mesogens or by a supermolecular aggregate or a dendrimer, . . .) (Guillon, 1999).

The molecular order in columnar phases can be described by the parameter ρ given by (Kats, 1978):

$$\rho = \rho_{\perp} f(x, y) + \chi \tag{2.11}$$

where $\rho_{\perp} f(x, y)$ describes the density variation in the plane perpendicular to the column's axis (it tends toward a constant at the transition from a columnar to a discotic nematic phase) and χ describes the density variation along the column's axis (χ is zero for disordered phases and varies periodically for ordered columnar phases).

The free-energy density that describes the elastic deformations in columnar phases is given by (Chandrasekhar and Ranganath, 1990; Kats, 1978):

$$F = \frac{B}{2}\left(\frac{\partial u_x}{\partial x} + \frac{\partial u_y}{\partial y}\right)^2 + \frac{D}{2}\left[\left(\frac{\partial u_x}{\partial x} - \frac{\partial u_y}{\partial y}\right)^2 + \left(\frac{\partial u_x}{\partial y} + \frac{\partial u_y}{\partial x}\right)^2\right] + \frac{K_3}{2}\left[\left(\frac{\partial^2 u_x}{\partial z^2}\right)^2 + \left(\frac{\partial^2 u_y}{\partial z^2}\right)^2\right] \tag{2.12}$$

where u_x and u_y represent displacements in the plane perpendicular to the columns' axes, B and D are constants that characterize the deformation of the 2D lattice in the plane perpendicular to the columns, and K_3 is the Frank constant associated to bending of the columns.

This brief description of LC phases is mostly valid for low-molecular-weight polymer LCs and dendrimer LCs. The key aspects referred are those more relevant to the NMR studies described in the book. Additional details will be given ahead whenever necessary.

Chapter 3

Molecular Structures of Liquid-Crystalline Dendrimers

3.1 Introduction

Liquid-crystalline (LC) dendrimers may be obtained by introducing mesogenic elements, similar to those described before (see Figs. 2.2, 2.3, and 2.5) into the dendrimer structure. Also in polymer liquid crystals, mesogenic units are included into the polymeric structure leading to partially ordered phases. In those systems, the rigid anisometric mesogenic units (e.g., calamitic or discotic) play the essential role on the ordering of the mesophases and the linking flexible chains remain essentially disordered, inducing some fluidity to the system (see Fig. 3.1). In main-chain LC polymers (see Fig. 3.1a) the mesogenic units are included sequentially into the polymer chain (meaning that the rigid units "interrupt" the flexible polymeric chain). In side-chain LC polymers, the mesogenic units are connected laterally with respect to the flexible chain (see Fig. 3.1b,c). In both cases (main chain and side chain, respectively), the rigid elements may be connected to the polymer chain either terminally ("end-on" LC polymers) or laterally ("side-on" LC polymers) using a convenient spacer, for instance, an alkyl

NMR of Liquid Crystal Dendrimers
Carlos R. Cruz, João L. Figueirinhas, and Pedro J. Sebastião

ISBN 978-981-4745-72-7 (Hardcover), 978-981-4745-73-4 (eBook)
www.panstanford.com

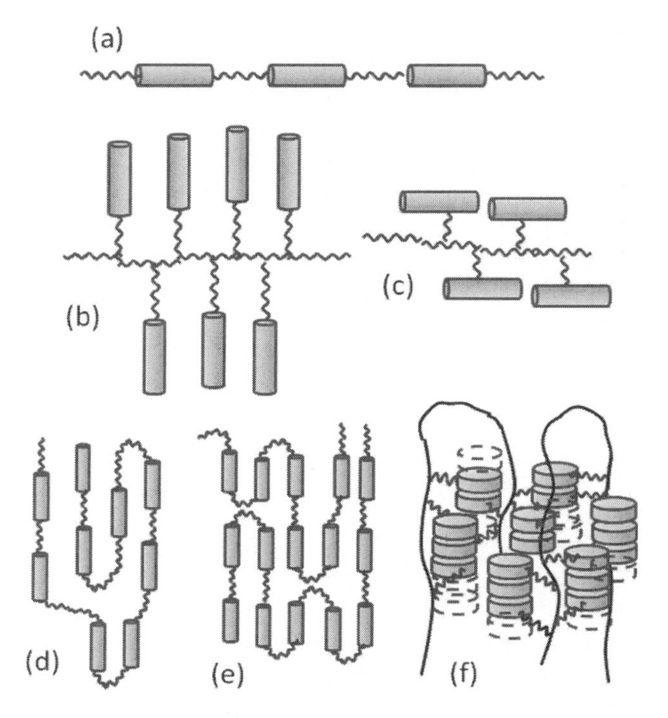

Figure 3.1 Schematic representation of different types of LC polymers and mesophases. (a) Main chain; (b) side chain, end-on; (c) side chain, side-on; (d) LC polymer nematic phase; (e) LC polymer smectic phase; and (f) LC polymer columnar phase.

chain or other flexible molecular segment. Figures 3.1b and 3.1c represent schematically these two cases, respectively. Figure 3.1d–f shows schematically some possible arrangements of polymeric chains and rigid mesogenic units corresponding to nematic, smectic, and columnar phases of LC polymers.

This very short and summarized description of the different types of LC polymers was introduced having in mind the analogy which can be established, to a certain extent, between LC polymers and LC dendrimers. Nevertheless, LC dendrimers present important additional features with respect to LC polymers concerning the design of mesogenic properties. Generally, in LC polymers, as referred above, the structure of the mesophase (nematic, smectic, columnar, . . .) is primarily defined by the mesogenic units (the chain essentially introduces some flexibility to the system). Differently, in LC dendrimers the dendritic structure plays a crucial role in the establishment of the mesomorphism.

Dendronized polymers are a particular type of side-chain polymers, where the lateral functional unit is a dendron or a dendrimer (Donnio et al., 2007). Those systems, which combine the structural defining characteristics of both polymers and dendrimers, represent a particular class of materials with complex molecular architecture and properties and were considered out of the scope of the present text.

As referred to in a previous chapter, dendrimers tend more and more to a globular shape as the generation increases. High-generation dendrimers are typically globular objects and therefore not the ideal candidates to generate LC phases. Contrary, lower generation dendrimers, especially if their core has some flexibility, may adopt different spatial configurations. If mesogenic units are included in the molecular structure of such systems, different mesophases may be obtained as the dendritic structure adopts convenient shapes in order to assure the common orientation of the mesogens.

The structure of a LC dendrimer depends on the chemical characteristic of the core, on the mesogenic units used to functionalize it and on the type of connection adopted (e.g., end-on versus side-on). The brief description that we present next is based on the classification by Donnio et al. in their extensive critical review on the subject (Donnio et al., 2007).

3.2 Side-Chain Liquid-Crystalline Dendrimers

The most studied of those systems are the so-called side-chain LC dendrimers. By analogy with polymers, side-chain LC dendrimers are composed of an inner dendritic core functionalized with specific mesogenic units in the outer shell.

As an example, Fig. 3.2 presents schematic pictures of side-chain generation G = 2 dendrimers with multiplicity of the core $N_C = 4$, and degree of branching $N_B = 2$, respectively, functionalized with

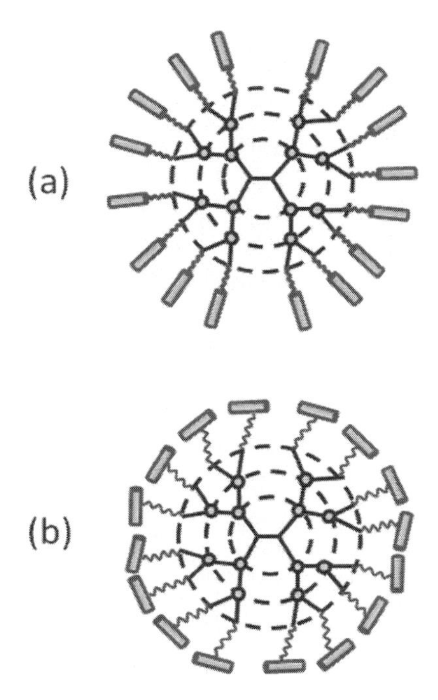

Figure 3.2 Schematic planar representation of generation 2 dendrimers with multiplicity of the core $N_C = 4$ and degree of branching $N_B = 2$. (a) Dendrimer functionalized with end-on mesogenic units; (b) dendrimer functionalized with side-on mesogenic units.

end-on (a) and side-on (b) mesogenic units. As shown in a previous chapter, the number of mesogenic units of a dendrimer molecule, in this particular case, is given by $Z = N_C N_B^G = 16$.

Considering applications, where the interaction of the dendrimers with external media (e.g., surfaces of confinement matrices) is crucial, the number and position of functional molecular moieties play an important role. In the case of side chain dendrimers, the molecule is functionalized on its external surface and the number of possible functional elements is determined by the number of terminal groups, Z, that can be tuned by selecting the appropriate dendrimer generation (see Eq. 1.1). Also, being attached to the dendrimer periphery, makes these functional elements more accessible to external media than in any other position in the dendritic structure. With respect to LC properties, mesogens positioned on the outside layer of a dendrimer interact closely to those belonging to neighboring molecules. The stabilization of LC dendrimer mesophases is determined by the competition between the tendency of the dendritic core to adopt an approximately isotropic conformation, and the tendency of average common orientation of mesogenic units. The energy–entropy balance results generally in the effect of microsegregation between molecular segments with different chemical properties, for example, aromatic rigid segments versus flexible aliphatic chains and dendritic core, in determined temperature ranges, contributing to the formation of mesophases (Tschierske, 1998). A resourceful combination of dendrimer core design and selection of mesogenic units used in the peripheral functionalization is a powerful tool to drive the synthesis of new molecules with specific mesogenic properties and potential applications. These interesting features of side-chain LC dendrimers justifies our special attention to this type of materials.

The side-chain LC dendrimers more extensively investigated in the literature are included on the two following types: (i) polyamidoamine (PAMAM)- and polypropyleneimine (PPI)-based LC dendrimers and (ii) silicon-containing LC dendrimers. These systems are particularly interesting due to the stability and versatility of their chemical structures with respect to potential functionalization and generation of mesomorphisms. Nuclear magnetic resonance (NMR) studies of LC dendrimers, to which this book is dedicated, focused essentially on these materials. The description of other systems like polyester and polyether LC dendrimers may also be found in (Donnio et al., 2007) and references therein.

3.2.1 PAMAM- and PPI-Based Liquid-Crystalline Dendrimers

PAMAM and PPI are amongst the most versatile dendritic structures, used in many applications, namely related to nanomedicine, for instance, as drug delivery platforms (Ward and Baker Jr., 2008). The structures of both PPI and PAMAM are based on a core with multiplicity $N_C = 4$ and a degree of branching $N_B = 2$. PAMAM's chemical structures of generations G0, G1, G2, and G3 are shown

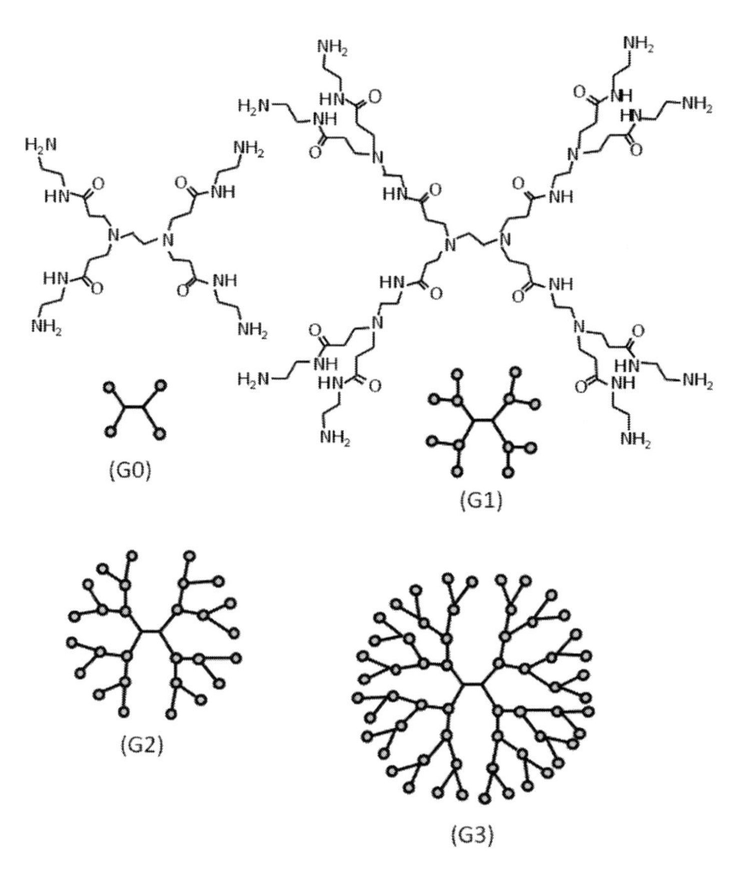

Figure 3.3 Chemical structures of PAMAM dendrimers of generations 0, 1, 2, and 3 (G2 and G3 schematic).

in Fig. 3.3 (G2 and G3 schematic). The difference between PAMAM and PPI is the inclusion of oxygen and nitrogen atoms in the linking segments of the dendrimer in the case of PAMAM. The $(CH_2)_2$-CONH-$(CH_2)_2$ segments in PAMAM are substituted by simply three CH_2 branching segments in the case of PPI. The properties of side-chain dendrimers based on PAMAM or PPI cores, in particular the induction of LC phases, depend on the appropriate choice of functional groups to attach to their terminal groups.

Groundbreaking works on PPI LC dendrimers were reported by Latterman et al. in 1997 (Cameron et al., 1997), and by Meijer at al. in 1998 (Baars et al., 1998), respectively. In the first case, the terminal functionalization of G0 to G4 PPI dendrimers with a nonmesogenic unit (a single benzene ring terminated by two flexible chains (CO(CH2)$_9$(CH3)) lead to the appearance of a hexagonal columnar phase in all cases. In the later, functionalization of dendrimers of generations G0, G2 and G4 with cyanobiphenyl terminal groups induces the formation of SmA phases.

Extensive and systematic investigations on PPI and PAMAM LC dendrimers have been carried out mainly by Serrano et al. at the University of Zaragoza and by Guillon, Donnio et al. at IPCMS-CNRS, Strasbourg (Barbera et al., 1999, 2001, 2005; Donnio et al., 2002; Marcos et al., 2001, 2003; Martin-Rapun et al., 2004; McKenna et al., 2005; Pastor et al., 2004; Rueff et al., 2003, 2006; Serrano et al., 2003). Very rich polymorphisms can be obtained in these systems through the functionalization of terminal branches of both PPI and PAMAM with appropriate mesogenic units. Different mesogens may be schematically defined as elongated aromatic segments terminated by one, two or three alkoxy chains, fan-shaped aromatic structures, or discotic (triphenylene) mesogens.

Dendrimers functionalized with mesogenic units bearing one terminal alkyl chain generally give rise to smectic phases due to their tendency to adopt an elongated overall shape as depicted in Fig. 3.4. In the case where mesogenic units with two or three terminal alkoxy chains are used, the formation of columnar phases is favored due to the tendency of the dendrimers to adopt a disc-like configuration.

These different configurations are guided by the tendency of microsegregation between dendritic cores, aromatic segments, and terminal alkoxy chains and by the steric constraints, which can be considered taking into account the area of the external dendrimer surface in the different cases. In the case of dendrimers terminated by mesogenic units with a single chain, the transverse area of the mesogenic sublayer in the molecule is similar to the cross-sectional area of the alkoxy chains, allowing for the approximated cylindrical conformation shown in Fig. 3.4a. This conformation favors the occurrence of the smectic phases (see Fig. 3.4c). In the case of two terminal alkoxy chains, the terminal cross-sectional area of the

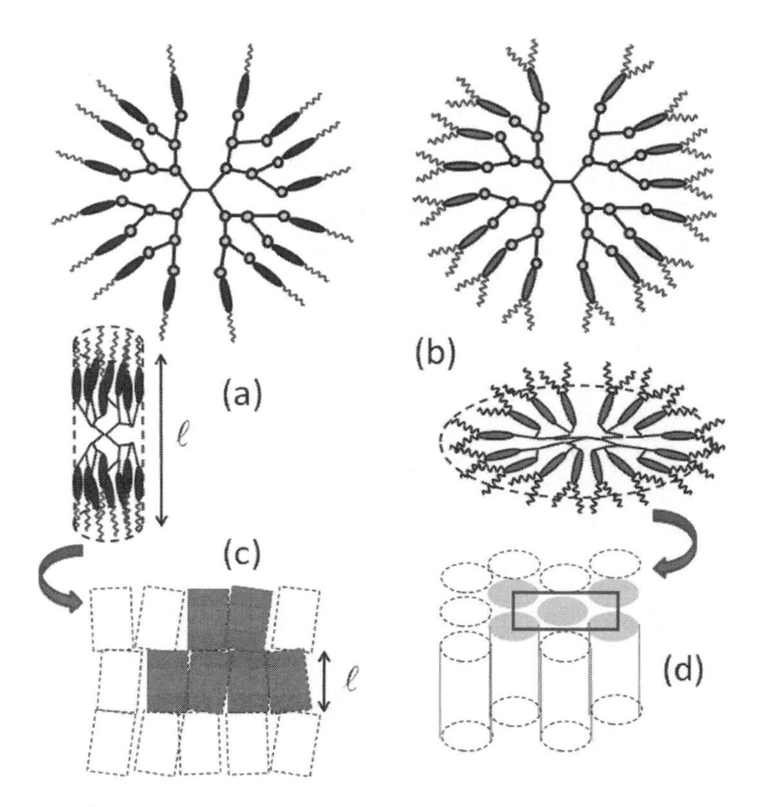

Figure 3.4 Schematic conformations of generation 2 PAMAM dendrimers end-on functionalized and respective molecular packing models: (a) dendrimer with mesogenic units bearing one terminal alkoxy chain, (b) dendrimer with mesogenic units bearing two terminal alkoxy chains, (c) structure of the SmA phase, and (d) structure of the columnar phase.

chains is clearly larger than that of the aromatic segments, inducing the disc-like conformation of the dendrimer (with the aromatic layer in the interior and the aliphatic chains at the external part of the disk as shown in Fig. 3.4b,d. In this case the microsegregation between the different molecular segments is achieved through the columnar phase structure: dendritic core in the central section of the columns, surrounded by the aromatic and aliphatic cylindrical layers, respectively.

Columnar phases are obtained by linking discotic (triphenylene) mesogenic units to PPI dendrimers of generations G1 to G5. A Col_r is observed for generation 1 dendrimers and Columnar Col_h are exhibited by generations G2–G5 (McKenna et al., 2005).

Nematic phases are also obtained with PPI dendrimers of generations G1–G5 side-on and end-on functionalized with elongated aromatic mesogenic units with short (C_2, C_4, C_5) alkoxy chains (Pastor et al., 2004). In the case of side-on dendrimers nematic phases are also obtained for longer linking alkoxy chains. The same effect can be produced using codendrimers partially functionalized with side-on and end-on mesogens. The coexistence between side-on and end-on segments is not enough to suppress the LC character of the system but in a given proportion is sufficient to avoid the effective microsegregation that would lead to the appearance of the smectic phase. The increasing of the proportion of end-on mesogens, with respect to the side-on part, suppresses the nematic phase and induces, SmA and SmC molecular arrangements (Martin-Rapun et al., 2004).

PAMAM dendrimers, using similar mesogenic units as those described for PPI, show, in general terms, similar mesogenic behaviour. However, due to the presence of CONH junctions in the dendritic branches, the mesophases obtained are, in general, thermodynamically more stable than those of PPI.

The synthesis of PAMAM codendrimers bearing fixed fractions of the total number of mesogenic units with one or two chains (randomly distributed at the dendritic periphery) gives rise to very interesting polymorphisms including SmA, SmC, Col_r and Col_h phases. The phase sequences of these systems can be tuned by selecting the relative proportion of both mesogenes in the co-dendrimer, taking into account that the one-chain mesogen favors the smectics and two-chain mesogenes favors the occurrence of columnar phases. For instance, if intermediate proportions of both mesogens are used, SmC and Col_r phases are formed between the SmA and Col_h exhibited by the corresponding homodendrimers.

A very rich variety of supermolecular ionic liquid crystals can also be obtained from PPI dendrimers functionalized with appropriate ionic moieties, combined with corresponding counterions (Cook et al., 2005; Martin-Rapun et al., 2005; Ramzi et al., 1999;

Tsiourvas et al., 2004; Ujiie et al., 2004). These systems exhibit SmA, different types of columnar phases and a cubic thermotropic phase (Tsiourvas et al., 2002). A hydrogen-bonding supermolecular complex dendrimer based on PPI, exhibiting SmA phases over a relatively broad temperature range from about 60°C to 140°C, was reported by Paleos et al. (Felekis et al., 2005).

3.2.2 Silicon-Containing Liquid-Crystalline Dendrimers

Most studies on silicon-containing LC dendrimers focus on three types of structures, regarding the chemical nature of the dendritic junctions: carbosilane (Si-C), siloxane (Si-O), and carbosilazane (Si-N).

3.2.2.1 Carbosilane-based dendrimers

The most extensive study on carbosilane-based dendrimers was carried out by Shibaev et al. (Agina et al., 2001; Klenin et al., 2001; Ponomarenko et al., 1999, 2000a,b, 2001a,b; Richardson et al., 1999a,b; Zhu et al., 2000). Dendrimers investigated in those works have a central core with $N_C = 4$, based on a tetravalent silicon atom, connected to a repetitive unit with a degree of branching $N_B = 2$, formed by a central silicon atom linked to three C_3 chains.

Dendrimers of this type (see Fig. 3.5a), show LC behavior from generation 1 (G1) up to 5 (G5) when functionalized with appropriate mesogenic units (see Fig. 3.5). Several examples are described in the literature and comprehensively listed in the work by Donnio et al. (2007) and references therein (Agina et al., 2001; Ponomarenko et al., 1996, 2001a). From G1 to G4, these dendrimers present uniquely SmA and SmC phases between room temperature and T about 90°C, when functionalized with calamitic units (e.g., cyanobiphenyl) terminally linked (end-on). It is interesting to notice that the mesogens used to functionalize the dendritic structure, in those studies, show nematic phases as pure monomers. The fact that the lamellar structure is induced by the presence of the dendritic core indicates the effect of a mechanism of microsegregation between the inner (flexible) core and the external rigid mesogenic functional elements. As in many other cases, the smectic

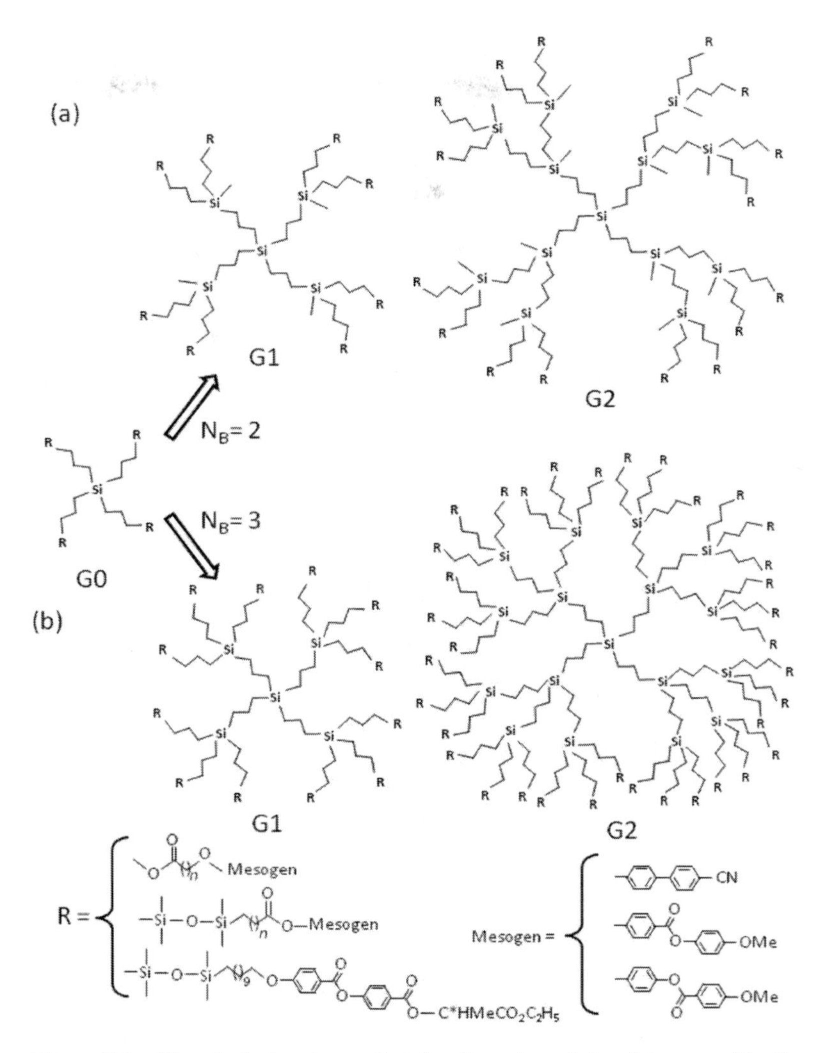

Figure 3.5 Chemical structure of carbosilane-based dendrimers with N_C = 4 and (a) N_B = 2 and (b) N_B = 3. The terminal groups, R, correspond to different types of aromatic mesogens terminally connected (end-on) by siloxane (Si-O) spacers to the dendrimer peripheral groups (Donnio et al., 2007).

phase is composed by alternating layers of dendritic cores and mesogenic external units. Interestingly, the smectic layers thickness is similar for the different generations, from G1 to G4. This is explained by a structural model where the dendritic core is deformed in the plane parallel to the smectic layers, assuming a conformation that allows for the mentioned microsegregation mechanism. In that way, typically, half of the mesogenic units in a dendrimer will remain parallel to each other, in a mesogenic sublayer, separated from the other half by the distorted dendritic core (see Fig. 3.6).

Besides smectic phases, generation 5 dendrimers of this type present also columnar phases. For increasing temperatures above the range corresponding to the SmA phase, first rectangular Col_r phases followed by hexagonal Col_h phases appear for the higher-generation dendrimers (G5). The authors explain this effect as a consequence of the tendency of the inner dendritic core to adopt increasingly expanded conformations, giving rise to the destabilization of the smectic layers as the temperature increases. The breaking of layers into columns occurs as the individual dendrimers change their overall shape form elongated cylinders to disk-like objects, driven by the tendency of the core to expand. In the columnar phases, according to the model, the dendrimers assume a shape similar to a disk with the mesogenic units radially disposed around the flattened dendritic core. In those phases, the dendrimers pile up, ensuring the microsegregation between the dendritic cores, which remain in the interior of the columns, and the mesogenic units that stay at the periphery (see Fig. 3.6). Lamellar-to-columnar phase transitions, where layers split into columns ensuring an effective microsegregation between molecular segments of different chemical characteristics were previously described in some low-molecular LC systems of biforked molecules (Guillon, 1999; Guillon et al., 1998).

The functionalization of this type of carbosilane dendrimers with specific chiral mesogenic groups lead to the formation of SmC* and SmA phases for lower generations (G1–G3). These systems exhibit switching properties, under the effect of alternating electric fields, typical of ferroelectric liquid crystals (Boiko et al., 2005; Zhu et al., 2000, 2001). Similar dendrimers of generations 4 and 5 exhibit a single Col_r phase. The grafting of a bent-core mesogenic unit to a

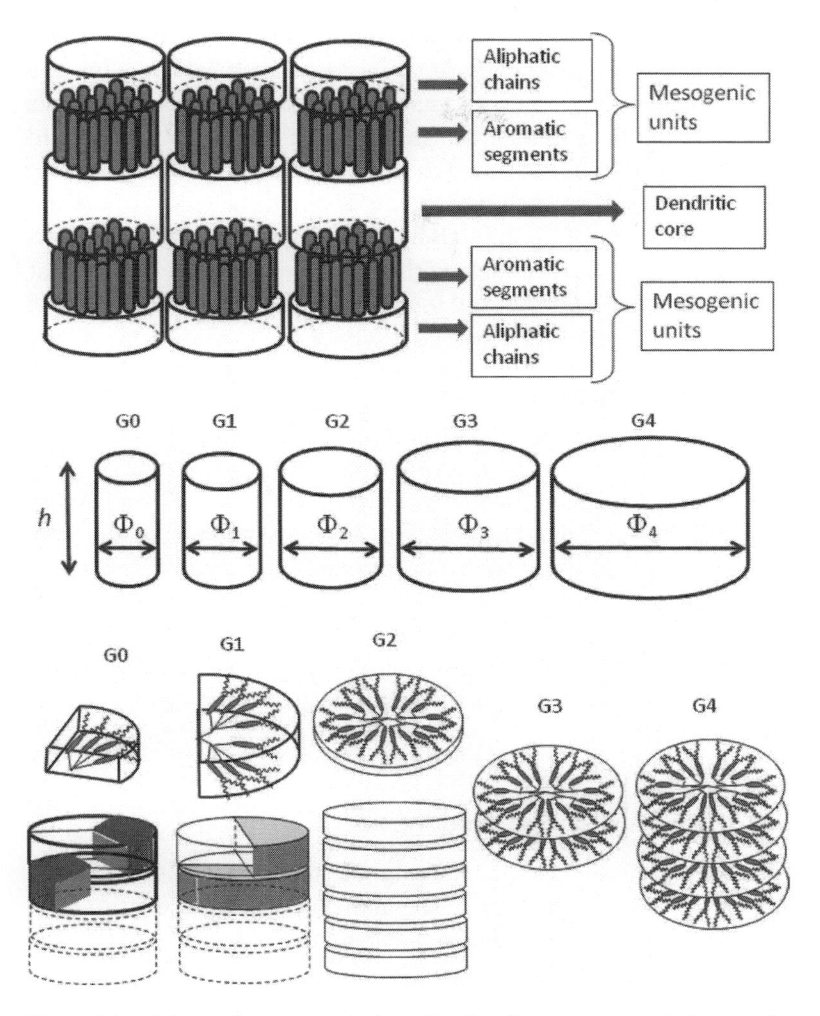

Figure 3.6 Schematic representation of molecular arrangements in smectic and columnar phases of dendrimers, showing the effect of microsegregation between the dendritic core and mesogenic units (Donnio and Guillon, 2006).

G2 dendrimer of this type gives rise to a new type of polar phase known as dark conglomerate phase. This novel mesophase is a special case of lamellar structure with layers composed by clusters with ferro- and antiferrolectric properties (Hahn et al., 2006). The functionalization of a G1 carbosilane dendrimer, as those described above, with an photosensitive azobenzene unit allowed for the study of these materials as candidates for optical data storage. This dendrimer exhibits a transition from SmA to isotropic liquid phase induced by UV irradiation as a consequence of the trans-cis isomerizaton of the azobenzene unit under the effect of radiation (Bobrovsky et al., 2001, 2002; Boiko et al., 2001).

Carbosilane dendrimers with the same central tetravalent described above, $N_C = 4$, and a more crowded dendritic structure, characterized by a branching multiplicity $N_B = 3$, functionalized with cholesteryl and cyanobiphenyl mesogenic groups (see Fig. 3.5b), exhibit smectic phases for generations 1 and 2. Higher generations functionalized with such mesogenic groups do not show any LC phases (Coen et al., 1996; Trahasch et al., 1999a,b). This fact can be explained by the increasing tendency of dendrimers with higher branching multiplicity to assume a globular shape with the increasing of the generation number. This characteristic is a consequence of the higher number of junctions in a dendrimer (expressed by variable J in Eq. 1.2 in the previous chapter). More junctions means more compact and less flexible dendritic core with less ability to allow for the self-organization of the peripheral mesogenic units.

3.2.2.2 Siloxane-based dendrimers

Another type of silicon-containing LC dendrimers is based on siloxane (Si-O). In this case, the most versatile core presented in the literature is a cage-like chemical structure (octa(dimethylsiloxy)octasilsesquioxane)), similar to a cube, with functional links at the eight vertices (see Fig. 3.7). This structure may be functionalized with siloxane branching units with ($N_B = 2$) as shown in Fig. 3.7, or with a single siloxane chain in each vertex forming a G0 dendrimer (octopode) (Donnio and Guillon, 2006; Elsasser et al., 2001, 2003; Mehl and Saez, 1999; Saez and Goodby, 1999, 2005; Saez et al., 2001). The G1 LC dendrimers are obtained through the functionalization of the 16 terminal branches with appropriate mesogenic units. Cyanobiphenyl or other

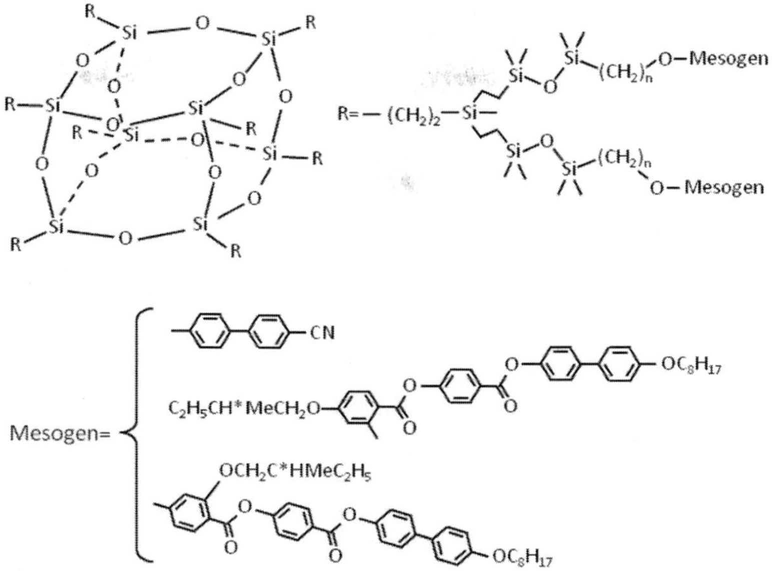

Figure 3.7 Chemical structure of octa(dimethylsiloxy) octasilsesquioxane used as a central core with (N_C = 8) of G0 and G1 siloxane containing dendrimers and functional linking units associated with G0 (octopodes) and G1 (N_B = 2) dendrimers (Donnio et al., 2007).

rigid elongated aromatic mesogenic units with different number of phenyl rings and terminal aliphatic chains can be used. End-on linking of the mesogens favors the formation of SmA and SmC phases, or SmC* if an appropriate chiral mesogen is used. Col_h and Col_r columnar phases appear in the case of side-on linked mesogens. Both in the case of SmC* and columnar phases a chiral nematic phase (N*) appears in a short temperature range. G0 dendrimers based on the silsesquioxane central unit (octopodes) functionalized with appropriate side-on mesogenic units give rise to a mesomorphism including a nematic phase and a columnar hexagonal phase for lower temperatures (Karahaliou et al., 2007, 2008). With respect to columnar phases, both in the case of G1 (with 16 terminal mesogens) and G0 (octopodes) with side-on mesogens, the silsesquioxane cubic structures form the inner core of the columns, which is surrounded by a region occupied by commonly oriented (in average) but positionally disordered mesogenic units. This molecular arrangement is effective for the microsegregation between the mesogenic units and the silsesquioxane dendritic cores. In Chapter 9 the structure of organosilaxane octopodes nematic and columnar phases are discussed in the framework of NMR relaxometry studies. These investigations relate the molecular organization of these dendrimers' LC phases with the respective molecular dynamics behavior.

Siloxane G0 dendrimers (tetrapodes) based on a tetravalent central silicon atom also exhibit very interesting polymorphisms, which include nematic, SmA, and SmC phases, depending on the selected mesogenic units and their linking position (end-on versus side-on) to the central core (Filip et al., 2007, 2010). In the case of siloxane tetrapodes functionalized with end-on cyanobiphenyl mesogenic units, the presence of the strong terminal dipole induces the formation of partial bilayered smectic phases SmA_d and SmC_d due to the competition between the tendency of microsegregation between aromatic, aliphatic, and siloxane sublayers, expressed by the smectic order parameter (see Eq. 2.8) and a antiferroelectric order parameter expressing the tendency of dipolar coupling between cyano groups belonging to different mesogens in adjacent layers (Filip et al., 2007). In the case of siloxane tetrapodes with side-on mesogens, nematic and SmC phases are shown. These materials are also remarkable due to the peculiar characteristics of the nematic phase, which they exhibit. This phase presents regions with SmC like local arrangement (cybotatic clusters) over the whole nematic domain, exhibiting nematic biaxial ordering (Cruz et al., 2008; Figueirinhas et al., 2005, 2009; Filip et al., 2010; Merkel et al., 2004; Polineni et al., 2013). This particular topic, with particular interest in LC Physics will be discussed in Chapter 8 dedicated to molecular ordering investigation using NMR. Molecular structures, phases sequences, and packing models of the LC phases exhibited by organosiloxane tetrapodes, both with end-on and side-on mesogenic units are presented in Chapters 8 and 9 dedicated to NMR spectroscopy and relaxometry experimental studies on LC dendrimers (Cardoso et al., 2008; Figueirinhas et al., 2005; Filip et al., 2007, 2010).

3.2.2.3 Carbosilazane-based dendrimers

Carbosilazane (Si-N)-based dendrimers is another example of silicon-containing dendrimers presented in the literature. Series of systems of this type, of generations G0, G1, and G2, based on a core with multiplicity (N_C = 3) and a branching degree of multiplicity (N_B = 2), functionalized with side-on mesogenic units were reported by Elsasser et al. (Elsasser et al., 2001, 2003). G0 generation dendrimers (tripods) present a single nematic phase at room temperature (Elsasser et al., 2001). The G1 systems present N and SmC phases (with decreasing temperatures (from 120°–130° to 50°–60°), and the G2 systems present a N-Col_h phases' sequence for similar temperature ranges.

From the description of many examples of side-chain LC dendrimers mentioned above, presented in the literature, it is possible to conclude that the functionalization of the dendritic core with laterally attached (side-on) elongated mesogenic units favors the occurrence of nematic phases. This is particularly effective for low generation dendrimers (typically G0, and G1).

On the other hand, terminally attached (end-on) mesogenic units lead to the appearance of smectic and/or columnar phases depending on the dendrimers generations and on the temperature range. Lower generation dendrimers, which usually correspond to more flexible shape cores tend to favor the formation of smectic phases. In this case, the mesogenic and the dendritic units become effectively microsegregated due to the stretching of the eventually easily deformable dendritic core, leading to the formation of submolecular layers of chemically distinct nature (dendritic core versus mesogenic units). Higher generations (with values depending on the particular characteristics of the system) and/or lower temperatures favor the formation of columnar phases as the dendritic core changes from an elongated to a disc-like configuration.

On the other hand, even for the side-on dendrimers, as the generation increases, the possibility of formation of nematic phases tend to disappear as the microsegregation between dendritic cores and mesogenic units becomes more effective, leading to smectic and columnar arrangements, especially for lower temperatures.

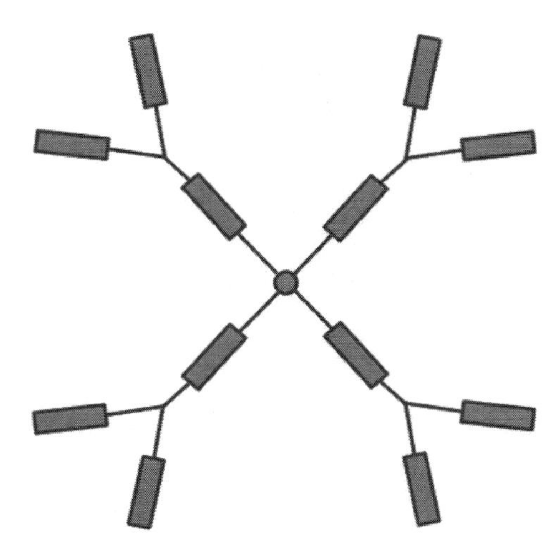

Figure 3.8 Schematic representation of a main-chain LC dendrimer of generation 1 with $N_C = 4$ and $N_B = 2$.

3.3 Main-Chain Liquid-Crystalline Dendrimers

Similarly to the synthetic strategy used in main-chain LC polymers, as discussed in the beginning of this section, LC properties can be induced in dendrimers through the inclusion of mesogenic units, sequentially attached by their extremes into the repeating dendritic branches (see Fig. 3.8).

In the cases of side-chain dendrimers described above, the dendritic branches are generally flexible giving rise to easily deformable structures, very well adapted to the formation of LC phases, especially for low generations. Considering that the branches of the main-chain dendrimers include some rigid moieties associated to the mesogenic elements, the use of a structure as shown in Fig. 3.8, induces less conformational freedom to the dendritic core. Given this restriction in flexibility, dendrimer branches lose their tendency to radiate isotropically, and, instead, they tend to give rise to anisometric structures depending on the choice of mesogenic units and on the steric properties of the dendritic structure.

When compared to side-chain dendrimers, main-chain dendrimers bear a much larger number of functional units for equivalent generations since they own one in every branch and not just at the periphery. The number of functional groups in a main chain dendrimer Z_{MC} will be significantly larger than that of a side chain dendrimer $Z_{\mathrm{SC}} = Z = N_{\mathrm{C}} N_{\mathrm{B}}^{G}$ given by Eq. 1.1.

$$Z_{\mathrm{MC}} = Z_{\mathrm{SC}} + J = N_{\mathrm{C}} N_{\mathrm{B}}^{G} + N_{\mathrm{C}} \left[\frac{(N_{\mathrm{B}}^{G} - 1)}{N_{\mathrm{B}} - 1} \right] = N_{\mathrm{C}} \left[\frac{(N_{\mathrm{B}}^{G+1} - 1)}{N_{\mathrm{B}} - 1} \right] \tag{3.1}$$

where the expression for the number of junctions J is given by Eq. 1.2. For instance, if the particular case of a G2 dendrimer with $N_{\mathrm{C}} = 4$ and $N_{\mathrm{B}} = 2$, with an additional generation to that schematically represented in Fig. 3.8, the number of functional elements will be

$$Z_{\mathrm{MC}} = 28 = 4 + 8 + 16 = 4 \left[\frac{(2^3 - 1)}{2 - 1} \right]$$

The increasing number of functional elements in the dendrimer has two potential effects (besides the discussed additional structural rigidity). On the one hand, eventual properties associated to the number of mesogens are expected to be enhanced. On the other hand, functional properties of the mesogens tend to be active not only at the dendrimer external surface (as in the case of side chain dendrimers) but all over the dendritic structure.

Two detailed studied examples of these systems reported in the literature are presented in Figs. 3.9 and 3.10. The so-called willow-like dendrimers, synthesized by Percec et al., which chemical structure is shown in Fig. 3.9, are based on triphenylene mesogenic units and present N and SmA phases for generations G1 to G3 (Li et al., 1996; Percec et al., 1995). The reasons for this mesomorphic behavior are explained by the possibility of the dendrimers' molecules to assume a rod-like overall shape rather than a disc-like one as the depicted chemical structure seems to indicate. This possibility results from the fact that the connection between the triphenylene and phenylene units presents conformational freedom that allows for a configuration where the phenylene unit and triphenylene units assume a overall shape, at given temperature ranges, compatible with the phases typical of calamitic mesogens.

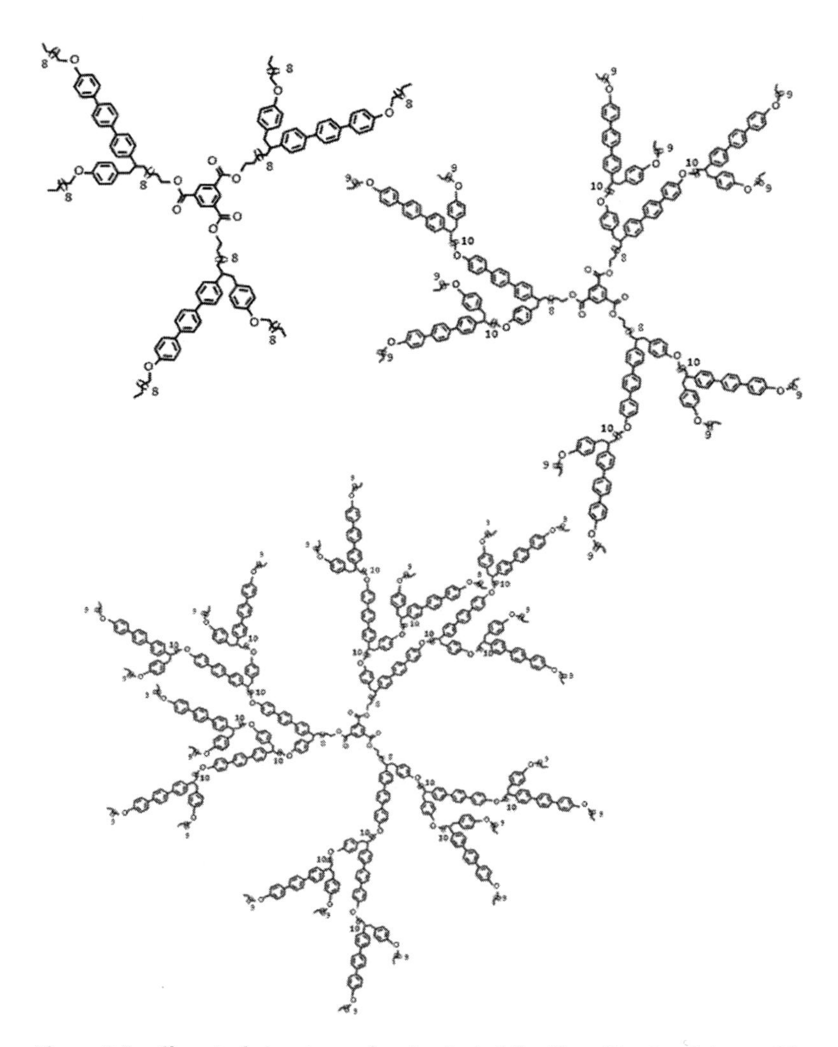

Figure 3.9 Chemical structure of main-chain LC willow-like dendrimers G0 to G2 (Donnio et al., 2007).

The mesomorphism of octopus dendrimers, whose chemical structure is shown in Fig. 3.10, depends on the number of terminal chains in the most external mesogenic units. In the case of a single terminal alkyl chain per mesogenic unit, the whole dendrimer tend to assume a global elongated shape giving rise to smectic

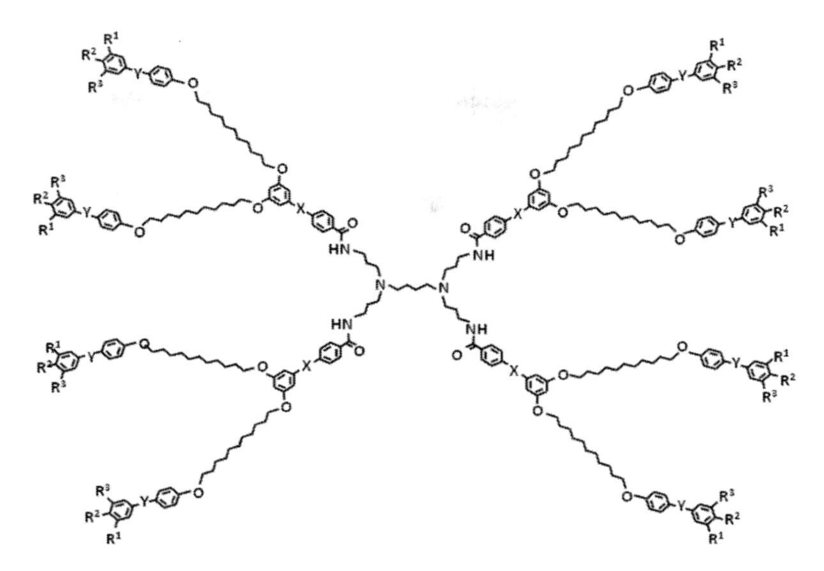

Figure 3.10 Chemical structure of main-chain LC octopus dendrimers, X, Y: $= / \equiv$; R^1, R^2, R^3 = $H/OC_{12}H_{25}$ (Donnio et al., 2007).

phases. In this case, as schematically represented in Fig. 3.11, the dendrimers assume a molecular organization where sublayers of the inner dendritic core alternate with sublayers of external dendritic mesogens, separated by aliphatic sublayers. In this arrangement, the mesogens of the layer composed by the inner dendritic core are forced to tilt in order to assume a compatible transverse molecular area per mesogenic unit similar to the one of the external dendritic layer, which has the double of the mesogenic units (see Fig. 3.11a). Detailed structural studies of this systems, investigated by Guillon, Donnio et al. in Strasbourg, show that the tilting directions of the rigid mesogenic cores belonging to the inner dendritic mesogenic layer are uncorrelated (Gehringer et al., 2003, 2004, 2005).

Dendrimers with more than one terminal aliphatic chain in the peripheral mesogenic units tend to assume a more spread shape giving rise to Col_h phases. In that case, individual dendrimers self-organize in such a way that the cross section of a column includes more than one dendritic molecule. These supramolecular arrangements are determined by the compatibility between the

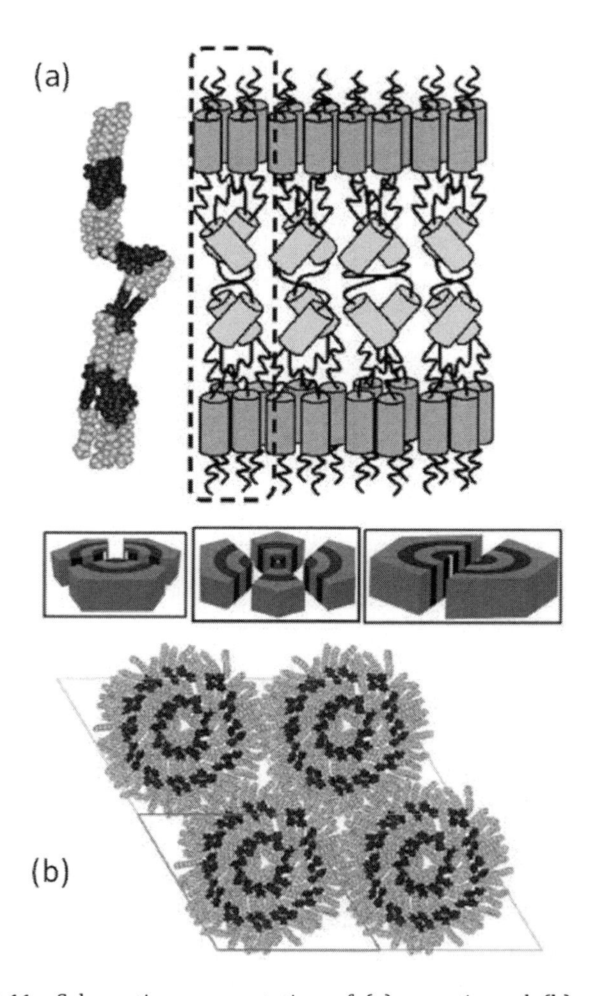

Figure 3.11 Schematic representation of (a) smectic and (b) columnar mesophases formed by octopus dendrimers. Reproduced from Donnio, B., Buathong, S., Bury, I. and Guillon, D. (2007). Liquid crystalline dendrimers, *Chemical Society Reviews* **36**, 9, pp. 1495–1513, with permission of The Royal Society of Chemistry.

cross-sectional areas of the different molecular segments of the dendrimer and the dimensions of the resulting columns. Details are given in references: (Gehringer et al., 2003, 2004, 2005) (see Fig. 3.11b).

3.4 Shape-Persistent Liquid-Crystalline Dendrimers

Contrary to side-chain and main-chain LC dendrimers, shape persistent LC dendrimers' molecules are formed by rigid segments only, without any linking flexible linkers. These dendrimers are completely rigid, assuming an intrinsic planar disc-like structure. When functionalized at the periphery with flexible aliphatic chains, these rigid cores' dendrimers give rise to columnar phases, as could be expected from their structure. An example of chemical structure of such materials is presented in Fig. 3.12 (Lehmann et al., 1999; Meier et al., 2000; Pesak and Moore, 1997). As in previous examples, the columnar structure allows for the effective microsegregation between the rigid dendritic cores, which form the central part of the columns, and the aliphatic chains that surround them, filling the intercolumnar space. The compounds presented in the literature show mesomorphic behavior for generations G1–G2 or G1–G3, respectively, depending on the characteristics of the terminal chains and linking groups X (see Fig. 3.12).

3.5 Supramolecular Dendromesogens

Supramolecular LC dendromesogens result from a powerful and extremely versatile synthetic strategy, introduced by Percec where dendrons of different generations and shapes converge into complex mesomorphic self-assembling supramolecular structures (Percec et al., 2004, 2007). This approach is akin to the convergent method of dendrimer synthesis, shortly described in the previous chapter. The elegance of this method lies on the contrast between the extreme simplicity of the dendritic building blocks (Percec dendrons) and the enormous variety of supramolecular mesomorphic structures generated through a judicious selection of small variations of those blocks. The general structure of the Percec dendritic units AB_n (where n refers to the branching multiplicity of a given single AB unit of the dendron) is based on an AB benzyl ether junction, generally functionalized at the apex by a given linking chemical group (X) and

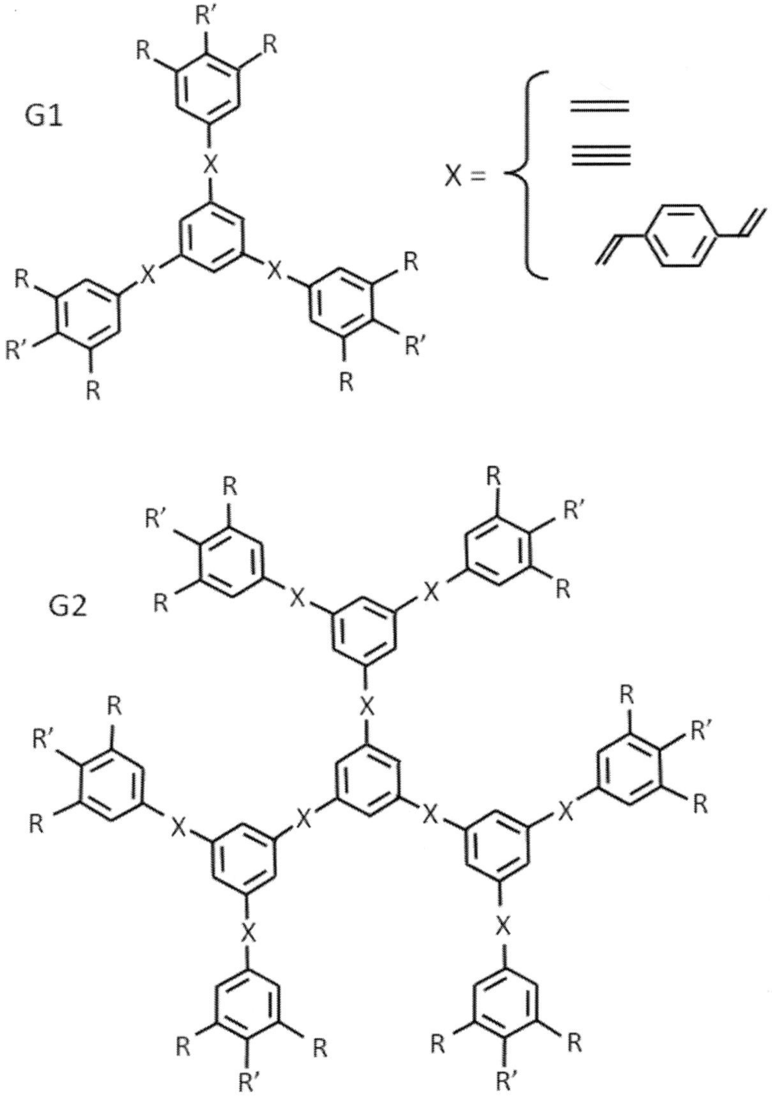

Figure 3.12 Chemical structure of shape-persistent LC dendrimer molecules of generations G1 and G2 (tolanoid: R=$CO_2(CH_2CH_2O)_3Me$, R′=H; stilbenoid: R=R′=OC_nH_{2n+1}, $n = 6, 12$) (Donnio et al., 2007).

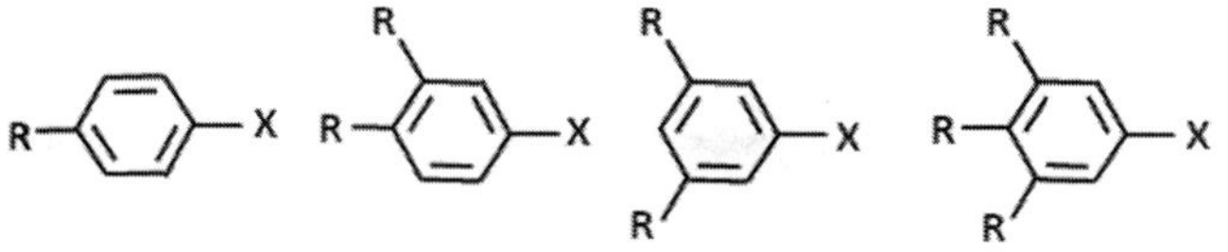

AB_n = 4-AB=AB 3,4-AB_2 3,5-AB_2 3,4,5-AB_3=AB_3
n = 1 2 2 3

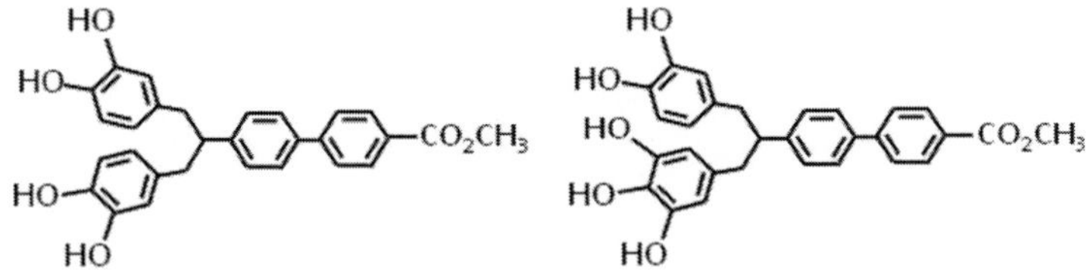

AB_n = AB_4-CO_2CH_3 AB_5-CO_2CH_3
n = 4 5

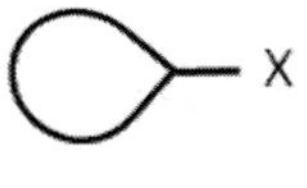

G1X=AB_nX

Figure 3.13 Chemical structure of Percec dendritic units AB_n with n = 1, 2, 3 (Percec et al., 2004) and $n = 4, 5$ (Percec et al., 2007).

at the periphery by flexible chains (R) (see Fig. 3.13). More complex structures allow for the design of AB_n units with $n = 4, 5$ (Percec et al., 2007).

Using dendritic sequences of units such as those represented in Fig. 3.13, complete "libraries" of dendrons with tuned shapes may be constructed. An example of those sequences and the related dendrons chemical structure is presented in Fig. 3.14 (Donnio et al., 2007).

The shape of the resulting dendrons, which have been systematically investigated from generations G1 to G5, depend on the linking unit X (examples: X=CO_2H, X=CO_2OH), on the number of peripheral chains n (which defines the branching multiplicity) and

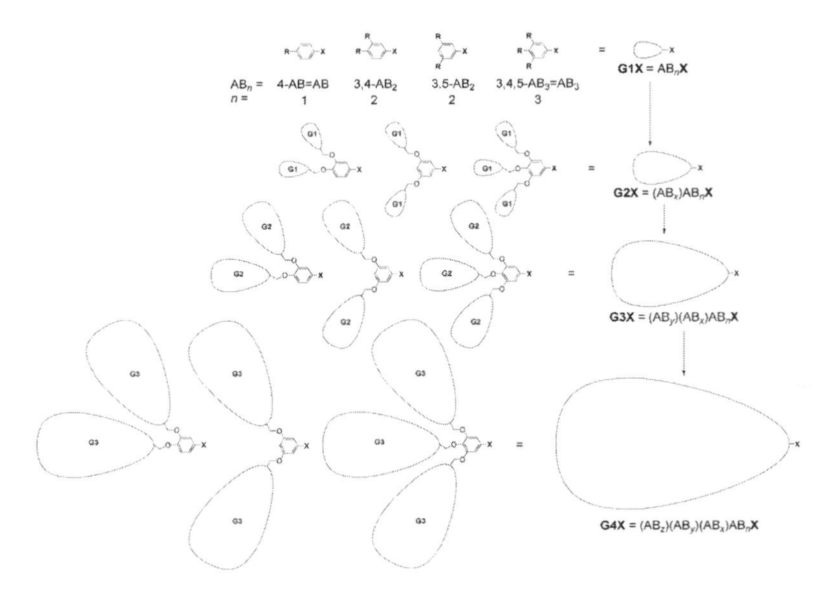

Figure 3.14 Schematic representation of Percec dendron structures of generations G1, G2, and G3. Reproduced from Donnio, B., Buathong, S., Bury, I. and Guillon, D. (2007). Liquid crystalline dendrimers, *Chemical Society Reviews* **36**, 9, pp. 1495–1513, with permission of The Royal Society of Chemistry.

their grafting position on the benzene ring. Different results can also be obtained through repetitive branching of similar dendritic units, formally represented by $(AB_n)^G X$, or by mixed arborescence, $(AB_x)(AB_y)(AB_z)(AB_n)X$ with $x, y, z, n = 1, 2, 3$ (see Fig. 3.14 for examples of different generations, branching multiplicities, and chains' positions).

This method allow for the design of dendrons of tuned size and shape, like fan shape, disc like, conical, hemispheric and spherical. The resulting monodendrons then self-assemble into supramolecular dendrimers due to microsegregation processes, steric constraints and hydrogen bonding. The resulting supramolecular objects self-organize generating a rich variety of mesophases. As a simple example it may be referred that the dendritic units with branching multiplicity $n = N_B = 2$ and branching flexible chains linked to positions 3 and 5 of the benzene ring (symmetric with

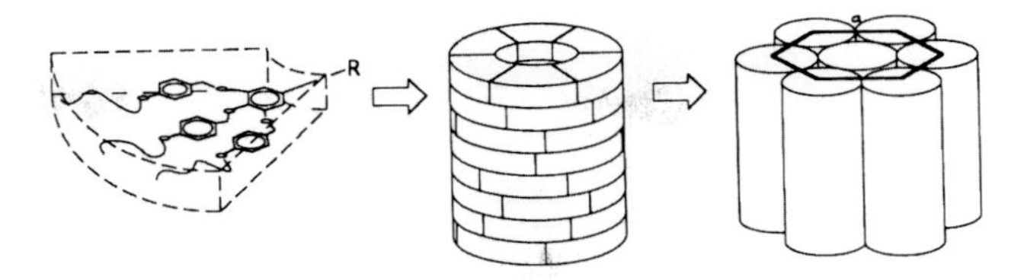

Figure 3.15 Schematic representation of columnar hexagonal (Col_h) phase resulting from tapered monodendrons $(AB)(3, 4, 5AB_3)$ self-assembling into discs that pile up into columns. From Donnio, B. and Guillon, D. (2006). Liquid crystalline dendrimers and polypedes, in *Supramolecular Polymers Polymeric Betains Oligomers*, Vol. 201 (Springer), pp. 45–155. With permission of Springer.

respect to the AB unit apex) tend to give rise to Col_h phases. In this case the monodendrons are approximately flat fan-shaped objects, which tend to organize into discs that pile up into columns (see Fig. 3.15). Higher branching multiplicities $n = 3$ tend to induce 3D dendritic growth that lead to conical, hemispherical or spherical objects that self-organize into 3D phases like micellar cubic phases or others.

In general terms, this type of systems give rise to smectic and columnar phases (Col_r, Col_h), different types of cubic mesophases (CubI, CubP), and other 3D mesophases (tetragonal and liquid quasi-crystal [LQC]). A scheme representing the formation of these LC systems is shown in Fig. 3.16.

More recently, an innovative strategy for the design of new libraries of self-assembling dendrons was proposed by Percec et al. (Rosen et al., 2010). That novel method is based on the deconstruction of previously known dendrons with different structures as an approach to the prediction of new complex self-organizing supramolecular architectures. According to the authors, this predictability leads to the outlining of a "nanoperiodic table" relating the structure of the dendritic building blocks with the properties of the resulting complex supramolecular architectures. The potential of such method of design of supramolecular self-organizing objects is exemplified by the description of synthetic supramolecular spheres composed by 933 quasiequivalent dendrons with a diameter of 17.2 nm, comparable in size with the most

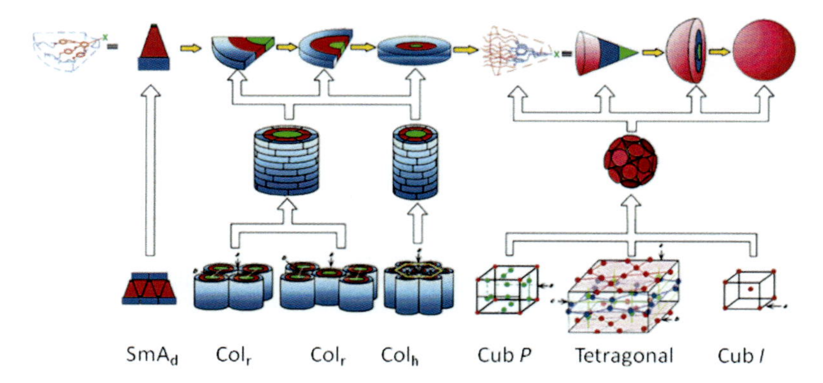

Figure 3.16 Schematic representation of the rich variety of mesomorphic structures generated by Percec LC supramolecular dendrimers. Reprinted with permission from Percec, V., Mitchell, C. M., Cho, W. D., Uchida, S., Glodde, M., Ungar, G., Zeng, X. B., Liu, Y. S., Balagurusamy, V. S. K. and Heiney, P. A. (2004). Designing libraries of first generation ab(3) and ab(2) self-assembling dendrons via the primary structure generated from combinations of (ab)(y)-ab(3) and (ab)(y)-ab(2) building blocks, *Journal of the American Chemical Society* **126**, 19, pp. 6078–6094. Copyright (2004), American Chemical Society.

complex biological structures such as ribosomes (Ramakrishnan and Moore, 2001).

3.6 Other Liquid-Crystalline Dendrimer Structures

Besides the previously described categories of LC dendrimers (side-chain, main-chain, shape-persistent, and supramolecular dendromesogens), there are other types, which specificity doesn't match the scope of anyone of the referred groups. Dendrimers containing active ferrocene and/or C60 units are an example of such systems, reported for the first time by Deschenaux et al. (Chuard et al., 2003; Dardel et al., 1999; Deschenaux et al., 1997, 2007; Lenoble et al., 2007). The interest of these types of systems is related to their potential applications as LC switches based on the photoinduced electron transfer, which occurs from ferrocene

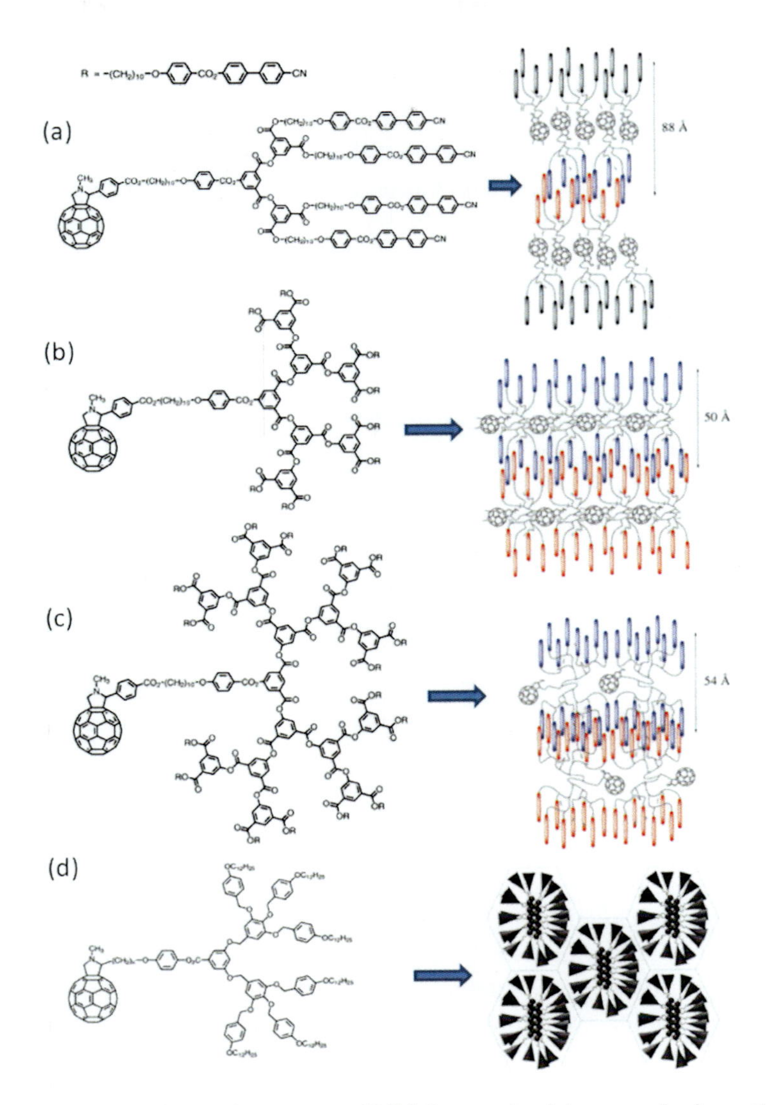

Figure 3.17 Chemical structure of LC fullerene dendrimers and schematic representation of SmA and Col_r phases exhibited by these systems: (a–c) SmA and (d) Col_r. Reproduced from Deschenaux, R., Donnio, B. and Guillon, D. (2007). Liquid-crystalline fullerodendrimers, *New Journal of Chemistry* **31**, 7, pp. 1064–1073, with permission of The Royal Society of Chemistry.

to fullerene, in the case where a system containing both chemical species is designed. Typically, most of the ferrocene and fullerene and ferrocene–fullerene LC dendrimers reported, with dendrimers functionalized with mesogenic units bearing terminal cyano groups at the periphery, exhibit SmA phases at high temperatures (typically >200°C for generations G1 to G4. In the case of G1 fullerene dendrimers, a bilayer structure results from segregation between C_{60} and the dendritic branches, where branches belonging to dendrimers on neighboring layers interact trough the respective terminal dipoles of the CN groups see Fig. 3.17a. From G2 to G4 the dendritic branches spread laterally, conditioning the C_{60} units to be completely involved by the dendritic branches (see Fig. 3.17b,c). Other particular cases where reported, of LC dendrimers of this type, exhibiting N* (Campidelli et al., 2003) and Col_r (Lenoble et al., 2006) phases, respectively (see Fig. 3.17d).

Metallodendromesogens are another interesting example of dendritic liquid crystals, which include metallic elements in their chemical composition. The organometallic character of these systems gives rise to important properties relevant to applications, such as electroactive molecules due to the multiredox centers, sensors due to their selective binding properties, and molecular antennas due to their photophysical properties (Gorman, 1998; Newkome et al., 1999). Different types of dendritic metallomesogens include purely covalent or ionic LC's, and both lyotropic (Donnio, 2002) and thermotropic (Donnio et al., 2003) systems. Generally, organometallic dendromesogens are supramolecular entities that incorporate chemical structures containing metallic atoms in the dendritic core from which standard organic dendritic branches (e.g., PPI) irradiate. Columnar phases, Col_h and Col_r are examples of mesophases exhibited by systems of this type (Stebani and Lattermann, 1995).

3.7 Summary

In this chapter we presented a general overview of molecular structures of LC dendrimers reported in the literature. LC dendrimers chemistry is an active research area, the design of new molecules is happening with some frequency and, therefore, the examples brought in here could not possibly represent the entire universe of such systems. However, more than looking for a kind of up-to-date list of LC dendritic compounds, the intention of this text is to provide material for reflection about general trends of the relation between molecular architectures and the structure of dendrimers' LC phases. First of all, besides some interesting peculiar LC dendrimers, briefly referred, including dendritic metallomesogens and fullerene containing dendrimers, two main categories were identified: (i) LC dendrimers based on covalent bonding and (ii) Supramolecular LC dendromesogens.

As mentioned above, the second type (ii) results from a synthetic strategy akin to convergent method defined in Chapter 1 and the LC dendrimers are supermolecular objects formed by dendrons that self-organize due to microsegregation processes, steric constraints and hydrogen bonding. The shape of the dendrons (which generation determines the size of the resulting dendrimer) is determined by the number of ramifications of the single building blocks (Percec dendron) and it varies from flat 2D shapes (elongated, fan-shaped, disk-like) to 3D shapes (cylindrical, flat tapered, conical, hemispherical). This procedure then gives rise to a rich variety of smectic, columnar, and 3D mesophases as described before, depending on the shape of the self-assembling dendrons.

In the first type (i) three subtypes were described, namely (a) side-chain LC dendrimers, (b) main-chain LC dendrimers, and (c) shape-persistent LC dendrimers. The two first subtypes (a) and (b) were defined by analogy with LC polymers. The third case (c) corresponds to dendrimers with a rigid structure, typically with a disc-like shape, where the LC character is induced by peripheral flexible chains. As mentioned above, these systems tend to exhibit columnar phases, resulting from the piling up of the individual dendrimers into columns. Main-chain LC dendrimers (b) are also relatively rigid objects (when compared to side-chain ones) due to the presence of rigid mesogenic units in each and all of the dendritic branches, but not so rigid as shape persistent LC dendrimers. In this case, some shape flexibility may still be induced by the dendritic links.

Side-chain LC dendrimers (a) are the most versatile and commonly presented in the literature. These systems can be obtained, in some cases, from standard dendrimers like PAMAM and PPI, by adequate functionalization of their peripheral chemical groups. In the most common cases, dendrimers for applications are prepared using this approach (Ward and Baker Jr., 2008). The formation of mesophases in side-chain LC dendrimers are driven by the competition between tendency of the flexible dendritic core, which tend to spread isotropically, and the peripheral mesogenic units that tend to mutually align allowing for the microsegregation of molecular segments with different chemical nature (dendritic core, aromatic rigid segments, terminal aliphatic chains, . . .). The flexibility of the dendritic core permit the dendrimers, especially those of lower generations (typically $G < G5$), to adopt different overall shapes, at different temperatures giving rise to smectic, columnar, and, in some cases, 3D LC phases.

Chapter 4

Fundamentals of Nuclear Magnetic Resonance

4.1 Introduction

Nuclear magnetic resonance (NMR) is a powerful analytical technique with application in materials science, biology, and medicine. Nowadays NMR is an essential tool for the synthetic chemist as well as the materials scientist, but it has made the connection with the great public through imagiology where magnetic resonance imaging (MRI) constitutes one of the most important methods of in vivo observation of structure and functionality of a living body. Foreseeing its great potential, the discovery and developments of this technique were recognized as major scientific achievements and have been rewarded with the attribution of several Nobel Prizes to those responsible for the discovery and development of the technique. It all started with the discovery of NMR in 1936 by Isidor Rabi while studying molecular beams (Rabi, 1937). Rabi found that a molecular beam exposed to a static magnetic field and submitted to radio waves can experience changes in the nuclear spins direction in a quantized form following quantum mechanics rules. For this discovery Rabi was awarded the Nobel Prize in Physics

NMR of Liquid Crystal Dendrimers
Carlos R. Cruz, João L. Figueirinhas, and Pedro J. Sebastião

ISBN 978-981-4745-72-7 (Hardcover), 978-981-4745-73-4 (eBook)
www.panstanford.com

in 1944. The extension of this technique to the study of liquids and solids by Felix Bloch and Edward M. Purcell in 1944 when they were trying to measure nuclear magnetic moments with higher precision using liquids and solids (Bloch, 1946; Purcell, 1946) granted them also the Nobel Prize in Physics that they shared in 1952. The first commercially available NMR spectrometer appeared in 1953 and since then a huge development of this technique took place. By 1970 pulsed methods were introduced in NMR and rapidly replaced the continuous-wave (CW) excitation (CW NMR) used up to then. The pulsed method, allied to the use of computer-implemented fast Fourier techniques, allowed the study of much more diluted samples and also much less abundant nuclear species. These pulsed techniques opened the way to what is called today 2D NMR where complex sequences of radio-frequency (RF) pulses generating 2D spectra are used. Richard R. Ernst, a prominent researcher in this area, was awarded the Nobel Prize in Chemistry "for his contributions to the development of the methodology of high-resolution nuclear magnetic resonance" in 1991 (Ernst, 1976). Since then, multidimensional NMR has grown continuously and in 2002 Kurt Wuthrich shared the Nobel Prize in Chemistry "for his development of nuclear magnetic resonance spectroscopy for determining the 3D structure of biological macromolecules in solution" (Wuthrich, 1983). As pointed before one very important development of the NMR technology that has reached the great public is MRI, an imaging technique with unique capabilities in the fields of anatomy and physiology, the relevance of this technique has been recognized and two leading researchers were distinguished in 2003, Paul C. Lauterbur and Sir Peter Mansfield were awarded the Nobel Prize in Physiology or Medicine "for their discoveries concerning magnetic resonance imaging."

The basis of NMR is the absorption and re-emission of electromagnetic energy in a discrete range of frequencies by a system of nuclear spins when appropriately excited. The target nucleus for studying NMR must possess a magnetic moment $\boldsymbol{\mu}$ and a spin angular momentum $\mathbf{S}$ which are related by a constant γ, the gyromagnetic ratio, characteristic of that nucleus, $\boldsymbol{\mu} = \gamma\mathbf{S} = \gamma\hbar\mathbf{I}$ where we have introduced the nucleus dimensionless angular momentum **I** and $\hbar$ is Planck's constant divided by 2π. Two types of information can be obtained from NMR experiments, (Abragam, 1961) one, of dynamic character, is extracted from the time evolution of macroscopic observables of the spin system after absorption of energy, the other, of static nature, is contained in the energy spectrum of the spin system. In the first kind of experiments, relaxation time measurements, the time evolution of the nuclear magnetization, after absorption of energy by the spin system is recorded. Several relaxation times can be determined; the most used ones are: T_1, spin–lattice relaxation time, T_2, spin–spin relaxation time, and $T_{1\rho}$ spin–lattice relaxation time in the rotating frame. These times are related to the correlation functions that describe the collective motions of the spins (Abragam, 1961). The second kind of experiments, absorption spectra studies, is concerned with the energy spectrum of the Spin Hamiltonian, they yield information on molecular structure and orientational order (Slichter, 1992).

4.2 Nuclear Paramagnetism

4.2.1 Nuclear Spin

Many nuclei are endowed with a magnetic moment and a spin angular momentum, the spin angular momentum is a vector and is quantized in both its norm and its projection along one spatial direction (Abragam, 1961). The spin quantum number I associated with the norm quantization can take either semi-integer or integer values. The nuclear spin I of an element nucleus with atomic number Z and mass number A takes the value 0 when both A and Z are even, takes an integer value when Z is odd and A is even, and takes a half-integer value when A is odd. The dimensionless spin angular momentum **I** is represented by a spin angular momentum operator **I**, in NMR experiments only the spin ground state is populated and the spin angular momentum norm squared given by the expectation value of $\mathbf{I}^2$ is calculated in terms of spin quantum number I by

$$|\mathbf{I}^2| = <\psi_i|\mathbf{I}^2|\psi_i> = I(I+1) \tag{4.1}$$

where $|\psi_i$ are the kets representing the wave functions of the spin ground state of the nucleus. Not only the norm of the spin angular momentum is quantized but also its component along a specified spatial direction, choosing the direction along the z axis one finds the expectation value of the z component of the spin angular momentum:

$$< \psi_i|I_z|\psi_i >= m_I \tag{4.2}$$

where m_I is the magnetic quantum number which can assume a discrete set of values:

$$m_I = \{-I, -I+1 \ldots I-1, I\}. \tag{4.3}$$

To each of the $2I+1$ values of m_I corresponds a different nuclear spin state described by a spin wave function $|\psi_i >$, the quantum numbers I and m_I define the spin state $|\psi_i >$ that can be written as $|I, m_I >$. These wave functions are eigenfunctions of both $\mathbf{I}^2$ and I_Z with eigenvalues $I(I+1)$ and m_I respectively. In the presence of a magnetic induction field **B**, a classical magnetic moment $\boldsymbol{\mu}$ acquires a potential energy given by

$$E = -\boldsymbol{\mu}.\mathbf{B}. \tag{4.4}$$

The corresponding Hamiltonian $\mathcal{H}$ for the nuclear moment in the presence of the static induction field $\mathbf{B} = B_z\mathbf{e}_z$ becomes

$$\mathcal{H} = -\hat{\mu}_z.B_z = -\gamma\hbar I_z B_z \tag{4.5}$$

which is known as the Zeeman Hamiltonian. Solving the eigenvalue equation for this Hamiltonian $\mathcal{H}|I, m_I >= E_{m_I}|I, m_I >$ leads to the set of energy levels:

$$E_{m_I} = -m_I\hbar\gamma B_z \tag{4.6}$$

and the energy difference between two adjacent levels becomes

$$\Delta E = E_{m_I} - E_{m_I+1} = \hbar\gamma B_z. \tag{4.7}$$

From this difference and using the Bohr condition for transitions between adjacent levels we obtain the frequency of the photon either absorbed or emitted when the spin angular momentum changes state with $\Delta m_I = -1$ or $+1$,

$$\nu = \frac{\Delta E}{h} = \frac{1}{2\pi}\gamma B_z \tag{4.8}$$

which is known as the Larmor precession frequency for reasons given further on. Because in NMR the directly observable state transitions are those for which $\Delta m_I = \pm 1$, this is the frequency of the electromagnetic fields that can be either absorbed or emitted by the spin system and thus plays a central role in NMR experiments. This frequency, as Eq. 4.8 shows, is set by the type of nuclei to be observed and the intensity of the external magnetic induction field used in the NMR experiment.

4.2.2 Nuclear Magnetization

When the spin system is subjected to an external magnetic induction field $\mathbf{B} = B_z\mathbf{e}_z$ and is in thermal equilibrium with the lattice, the different energy levels of the spin Hamiltonian become populated following a Boltzmann distribution and this leads to the appearance of a nuclear magnetization $\mathbf{M}$ along z which for a sample of N spins per unit of volume takes the value

$$\mathbf{M} = N\hbar\gamma \frac{\sum_{m=-I}^{I} m e^{\hbar\gamma m B_z/kT}}{\sum_{m=-I}^{I} e^{\hbar\gamma m B_z/kT}} \mathbf{e}_z. \tag{4.9}$$

Because the ratio $\hbar\gamma B_z/kT$ is in general a small number in nuclear magnetism, the exponentials in Eq. 4.9 can be approximated ($e^x \cong 1 + x$) leading to

$$M_z = \frac{N\hbar^2\gamma^2 B_z}{kT} \frac{\sum_{m=-I}^{I} m^2}{2I+1} = \frac{N\hbar^2\gamma^2 \mu_0 H_z I(I+1)}{3kT} = \chi H_z \tag{4.10}$$

where χ is the static nuclear magnetic susceptibility displaying the 1/T dependence characteristic of the Curie law of magnetism (Abragam, 1961). Equation 4.10 shows that an ensemble of nuclear spins in thermal equilibrium with the lattice develops a nuclear magnetization along the field proportional to the field strength. Later on it will be shown that this magnetization can be manipulated and its evolution back to equilibrium contains the relevant information on the spin system and its interaction with the lattice.

4.2.3 Interaction with an External Field

Interaction of the spin system with an oscillating magnetic induction field meeting the Bohr condition may lead to the absorption or radiation of energy by the spin system. To find the necessary conditions for emission/absorption processes to occur related to the state populations we can start by analyzing the simple case of an ensemble of noninteracting spin 1/2 nuclei under a static magnetic induction. Calling the populations of the two spin states respectively n_+ and n_- we can write the rate of variation of these populations considering that the interaction with the oscillating field produces identical rates of transition from the $(-)$ to the $(+)$ state and from the $(+)$ to $(-)$ state, $W_{(-)->(+)} = W_{(+)->(-)} = W$ (Slichter, 1992), leading to

$$\frac{dn_+}{dt} = W_{(-)(+)}n_- - W_{(+)(-)}n_+ = W(n_- - n_+)$$
$$\frac{dn_-}{dt} = W_{(+)(-)}n_+ - W_{(-)(+)}n_- = W(n_+ - n_-). \qquad (4.11)$$

Introducing now the two new variables $\Delta n = n_+ - n_-$ and $N = n_+ + n_-$, Eq. 4.11 convert to

$$\frac{d\Delta n}{dt} = -2W\Delta n, \frac{dN}{dt} = 0. \qquad (4.12)$$

whose solution is $N(t) = N(t = 0)$ and $\Delta n(t) = \Delta n(t = 0)e^{-2Wt}$. Considering n_- the population of states with $m = -1/2$, absorption of energy by the spin system occurs in the transition from the $m = 1/2$ to the $m = -1/2$ states, the rate of energy absorption by unit of time, dE/dt can be found as

$$\frac{dE}{dt} = Wn_+\hbar\omega - Wn_-\hbar\omega = W\Delta n\hbar\omega. \qquad (4.13)$$

For energy absorption to occur, dE/dt must be positive and this implies a positive population difference Δn which is found when the spin system is in thermal equilibrium with the lattice. Equation 4.12 shows us also that on the continuous application of the oscillating magnetic induction field the population difference decays exponentially to zero reaching what is called the saturation of the spin system. When this is achieved the system of Eq. 4.12 predicts no further evolution, even if the oscillation magnetic field is removed, which amounts to put $W = 0$. This is a shortcoming of this simple model that fully neglects the always present interaction of the spins with the lattice, which is responsible for bringing the spin system into thermal equilibrium with the lattice. In a real system one observes that in saturation the spin system ceases to absorb energy, but when the oscillating magnetic induction field is removed, the spin system relaxes back to thermal equilibrium with the lattice.

4.3 Dynamics of Noninteracting Spins: Bloch's Equations

4.3.1 Motion of a Magnetic Moment

In a classical description of the dynamic behavior of the spin system we start by considering a magnetic moment $\boldsymbol{\mu}$ in the presence of a static magnetic induction field oriented along the z axis of a laboratory fixed frame $\{x, y, z\}$. The magnetic moment becomes subjected to a torque amounting to $\boldsymbol{\mu} \times \mathbf{B}$ and the rate of change of its' angular momentum $\mathbf{S}$ becomes

$$\frac{d\mathbf{S}}{dt} = \boldsymbol{\mu} \times \mathbf{B} \tag{4.14}$$

and as $\boldsymbol{\mu} = \gamma \mathbf{S}$ we get

$$\frac{d\boldsymbol{\mu}}{dt} = \gamma \boldsymbol{\mu} \times \mathbf{B}. \tag{4.15}$$

To get an insight on the solution of Eq. 4.15 we consider a reference frame $\{x', y', z'\}$ with its z' axis coinciding with the z axis of the laboratory fixed frame but rotating around the z axis with a uniform angular velocity $\Omega = \Omega_z \mathbf{e}_z$. In this rotating frame, the rate of change of the magnetic moment $\boldsymbol{\mu}$ becomes

$$\frac{\delta \boldsymbol{\mu}}{\delta t} + \Omega \times \boldsymbol{\mu} = \gamma \boldsymbol{\mu} \times \mathbf{B}. \tag{4.16}$$

leading to

$$\frac{\delta \boldsymbol{\mu}}{\delta t} = \boldsymbol{\mu} \times (\gamma \mathbf{B} + \boldsymbol{\Omega}). \tag{4.17}$$

when $\Omega = \omega_0 = -\gamma \mathbf{B}$ one obtains $\frac{\delta \boldsymbol{\mu}}{\delta t} = 0$ and this means that $\boldsymbol{\mu}$ does not change in the rotating frame $\{x', y', z'\}$ or equivalently that $\boldsymbol{\mu}$ rotates in the laboratory fixed frame $\{x, y, z\}$ with angular velocity $\Omega = -\gamma \mathbf{B}$ around the static induction field, this rotation frequency is called the Larmor precession frequency and coincides with the photon frequency involved in the $\Delta m_I = \pm 1$ transitions referred above.

4.3.2 Bloch's Equations

In 1946 Felix Bloch (Bloch, 1946) proposed a set of phenomenological equations to describe the magnetic properties of ensembles of nuclei in external magnetic fields. These equations giving the time rate of change of the nuclear magnetization achieve in many cases with emphasis on isotropic liquids a correct description of the phenomenon. When the static part of **B** is along the laboratory fixed z axis the equations take the form

$$\frac{d\mathbf{M}}{dt} = \gamma \mathbf{M} \times \mathbf{B} - \frac{M_x \mathbf{e}_x + M_y \mathbf{e}_y}{T_2} - \frac{M_z - M_0}{T_1} \mathbf{e}_z. \tag{4.18}$$

Equation 4.18 indicates that three terms determine the time evolution of the nuclear magnetization **M**, the first term represents an interaction between the nuclear magnetization and the external magnetic induction **B** and promotes the precession of **M** around **B** with an angular frequency $\omega_0 = -\gamma B_z$, the second term promotes a decay toward 0 of the transverse components of the **M**, M_x and M_y, with a time constant T_2 which is name the spin–spin relaxation time as it describes the rate at which the spins exchange energy between them. The third term in the second member of Eq. 4.18 induces a decay of the z component of the magnetization toward its equilibrium value M_0. The time rate associated with this process, T_1 is named spin–lattice relaxation time as the decay of M_z corresponds to the exchange of energy between the spin system and the lattice. To account for the different terms in the second member of Eq. 4.18 we can start by considering an ensemble of N noninteracting spins per unit of volume and use Eq. 4.15 to obtain the rate of change of the total magnetic moment per unit of volume or magnetization as $\frac{d\mathbf{M}}{dt} = \gamma \mathbf{M} \times \mathbf{B}$ where the term in the second member is just the first term in the second member of Eq. 4.18. The other terms of Eq. 4.18 where introduced to account for the decay of the different components of the magnetization toward their equilibrium values of 0 for the transverse components of $\mathbf{M}$ and M_0 for its z component. The decay of M_z involves the transfer of energy between the spin system and the lattice and is governed by T_1 the spin–lattice relaxation time while the decays of M_x and M_y are determined by the exchange of energy between the different spins and are governed by T_2 the spin–spin relaxation time.

4.3.3 Interaction with an Oscillating Magnetic Induction

We next use Bloch's equations to investigate the interaction of an oscillating magnetic field with the spin system. The magnetic induction considered is the sum of a static part oriented along the z axis of the laboratory fixed frame $\{x, y, z\}$ and an oscillating part oriented along the x axis, $\mathbf{B} = B_0\mathbf{e}_z + B_1 cos(\omega t)\mathbf{e}_x$. The orientation of the oscillating part is chosen perpendicular to the static part to allow for energy transfer between the field and the spin system to occur. It is convenient for solving Bloch's equations to introduce a rotating frame $\left\{x', y', z'\right\}$ with its z' axis coinciding with the z axis of the fixed laboratory frame and rotating with angular velocity $\omega = \omega\mathbf{e}_{z'}$ around the z axis of the fixed frame. In the rotating frame Bloch's equations take the form

$$\frac{d\mathbf{M}'}{dt} = \gamma\mathbf{M}' \times \left(\mathbf{B}' + \frac{1}{\gamma}\omega\mathbf{e}_{\mathbf{z}'}\right) - \frac{M_{x'}\mathbf{e}_{x'} + M_{y'}\mathbf{e}_{y'}}{T_2} - \frac{M_{z'} - M_0}{T_1}\mathbf{e}_{z'}. \tag{4.19}$$

In the rotating frame the magnetic induction field $\mathbf{B}'$ is given by $\mathbf{B}' = B_0\mathbf{e}_{z'} + \frac{B_1}{2}\mathbf{e}_{x'} + \frac{B_1}{2}\left(cos(2\omega t)\mathbf{e}_{\mathbf{x}'} - sin(2\omega t)\mathbf{e}_{\mathbf{y}'}\right)$. When $\omega = \omega_0 = -\gamma B_0$ the term $\frac{\omega\mathbf{e}_{z'}}{\gamma}$ exactly cancels the term in $\mathbf{B}'$ due to the static magnetic induction, $B_0\mathbf{e}_{z'}$, and the system is said to be in resonance. The other two terms in $\mathbf{B}'$ represent a static contribution of amplitude $\frac{B_1}{2}$ along the x' axis and a rotating contribution with the same amplitude and angular frequency -2ω. This rotating component of the magnetic induction field in the rotating frame is not effective in inducing magnetization changes when compared with the static part oriented along the x' axis and is usually neglected. Bloch's equations in the rotating frame at resonance become

$$\frac{dM_{x'}}{dt} = -\frac{M_{x'}}{T_2}$$
$$\frac{dM_{y'}}{dt} = \gamma M_{z'}\frac{B_1}{2} - \frac{M_{y'}}{T_2}$$
$$\frac{dM_{z'}}{dt} = -\gamma M_{y'}\frac{B_1}{2} - \frac{M_{z'} - M_0}{T_1}. \quad (4.20)$$

An approximate solution for the system of Eq. 4.20 can be easily obtained for short time intervals and when $\gamma B_1 >> 1/T_1, 1/T_2$, in this case the relaxation terms in the Eqs. 4.20 can be neglected and the solution is:

$$M_{x'}(t) = 0$$
$$M_{y'}(t) = -M_0 \sin(\omega_1 t)$$
$$M_{z'}(t) = M_0 \cos(\omega_1 t) \quad (4.21)$$

where the initial condition $M(0) = M_0\boldsymbol{e_{z'}}$ was considered and $\omega_1 = -\gamma B_1/2$. The solution, Eq. 4.21, corresponds to a magnetization in the $\{y', z'\}$ plane rotating around the x' axis with angular velocity ω_1. If the B_1 field is active for a time interval $\tau = \pi/(\gamma B_1)$, the magnetization is rotated from the z' axis to the y' axis and this is called a $\pi/2$ pulse because the magnetization was rotated by $\pi/2$. If B_1 is active a time interval twice as long (2τ) then the magnetization is rotated by π around x' and this is called a π pulse. After the application of a $\pi/2$ RF pulse with B_1 along x' at resonance the magnetization **M** lies along the y' axis, its evolution to equilibrium can be found solving Eqs. 4.20 with $B_1 = 0$ leading to

$$M_{x'}(t) = 0$$
$$M_{y'}(t) = M_0 e^{\frac{-t}{T_2}}$$
$$M_{z'}(t) = M_0(1 - e^{\frac{-t}{T_1}}) \quad (4.22)$$

these results show that $M_{y'}$ decays exponentially to zero with time constant T_2 and $M_{z'}$ approaches is equilibrium value M_0 with time constant T_1.

4.3.4 Generating B_1 and Reading $M_\perp$

In high-field NMR experiments the nuclear magnetization **M** is manipulated with the application of short B_1 pulses at resonance, these pulses are created by a conducting coil oriented with its axis orthogonal to the static magnetic induction field and energized with an alternating current with angular frequency equal to the Larmor frequency. When these pulses are not being generated, the same coil is used to detect the time evolution of the magnetization and due to the coil's orientation it will detect the changes in the x' and y' components of the magnetization.

4.4 The Nuclear Spin Hamiltonian

The total Hamiltonian of a physical system is in general of prohibitive complexity for any attempt of finding the equations of motion for the full quantum mechanical system. The possibility of studying a simplified nuclear spin Hamiltonian of that system achieved by NMR is of great advantage. The nuclear spin Hamiltonian includes only nuclear spin operators and some phenomenological constants arising from the averaging of the other degrees of freedom of the total Hamiltonian. The nuclear spin Hamiltonian (Ernst et al., 1992) is composed of at most six terms that include the Zeeman interaction, $\mathcal{H}_Z$, between the nuclear spins and the external static magnetic induction, the interaction with the high (radio)-frequency magnetic induction field $\mathcal{H}_{\text{r.f.}}$, the chemical shift $\mathcal{H}_\sigma$, accounting for the field-shielding effect produced by the electronic clouds, the indirect spin–spin coupling $\mathcal{H}_J$, accounting for the internuclei interactions mediated by the surrounding electronic clouds, the direct dipolar interaction $\mathcal{H}_D$, and the quadrupolar interaction $\mathcal{H}_Q$, present for spins with $I \geq 1$ and accounting for the interaction between the electric quadrupole moment of the nucleus and the electric field gradient at the nucleus site. These different terms take the form

$$\mathcal{H}_Z = -\sum_{k=1}^{N} \gamma_k \hbar \mathbf{I}_k \mathbf{B}_0$$

$$\mathcal{H}_{\text{r.f.}} = -\sum_{k=1}^{N} \gamma_k \hbar \mathbf{I}_k \mathbf{B}_{\text{r.f.}}(t)$$

$$\mathcal{H}_\sigma = \sum_{k=1}^{N} \gamma_k \hbar \mathbf{I}_k \sigma_k \mathbf{B}_0$$

$$\mathcal{H}_J = 2\pi\hbar \sum_{k=1,l>k}^{N-1,N} \mathbf{I}_k \mathbf{J}_{kl} \mathbf{I}_l$$

$$\mathcal{H}_D = \sum_{k=1,l>k}^{N-1,N} \mathbf{I}_k \mathbf{D}_{kl} \mathbf{I}_l = \sum_{k=1,l>k}^{N-1,N} \frac{\mu_0 \gamma_k \gamma_l \hbar^2}{4\pi r_{kl}^3} \left\{ \mathbf{I}_k \mathbf{I}_l - \frac{3}{r_{kl}^2} (\mathbf{I}_k \mathbf{r}_{kl})(\mathbf{I}_l \mathbf{r}_{kl}) \right\}$$

$$\mathcal{H}_Q = \sum_{k=1}^{N} \mathbf{I}_k \mathbf{Q}_k \mathbf{I}_k = \sum_{k=1}^{N} \mathbf{I}_k \frac{eQ_k}{2I_k(2I_k - 1)} \mathbf{V}_k \mathbf{I}_k \tag{4.23}$$

where k and l run over all the N spins in the system and γ_k and $\mathbf{I}_k$ were seen before. The chemical-shielding tensors $\boldsymbol{\sigma}_k$, the indirect spin–spin coupling tensors $\mathbf{J}_{kl}$, the direct dipolar interaction tensors $\mathbf{D}_{kl}$, and the quadrupolar coupling tensors $\mathbf{Q}_k$ are quantities characteristic of the particular spin system that convey a significant part of the information NMR can obtain regarding the physical system the spins are included in. $\mathbf{r}_{kl}$ is the vector joining spins k and l, Q_k is the nuclear quadrupole moment of nucleus k and $\mathbf{V}_k$ is the electric field gradient tensor at the site of nucleus k.

4.4.1 Reduced Spin Hamiltonian in High-Field NMR Spectroscopy

In high-field NMR the Zeeman term $\mathcal{H}_Z$ dominates the spin Hamiltonian and in the absence of the high-frequency $\mathbf{B}_{\text{r.f.}}$ field, the energy levels of the total Hamiltonian can be determined using first-order perturbation theory, considering $\mathcal{H}_Z$ as the unperturbed Hamiltonian $\mathcal{H}_0 = \mathcal{H}_Z$ and the remaining interactions as the perturbation $\mathcal{H}_1 = \mathcal{H}_\sigma + \mathcal{H}_J + \mathcal{H}_D + \mathcal{H}_Q$. In this case the different terms in the perturbation Hamiltonian $\mathcal{H}_1$ can be simplified, remaining only those parts that contribute in first-order perturbation theory thus commuting with $\mathcal{H}_0$, they are called the adiabatic or secular part of $\mathcal{H}_1$ (Ernst et al., 1992). To find the secular part of $\mathcal{H}_1$ we note that all the terms in $\mathcal{H}_1$ correspond to a contraction of two second-rank tensors, one a tensor operator derived from the spin operators $S_{ij,lm} = I_{i,l} I_{j,m}$ and another tensor $R^{\alpha}_{ij,lm}$ dependent upon the interaction considered α:

$$\mathcal{H}_{1,\alpha} = \sum_{i=1,\,j>i}^{N-1,N} \sum_{l=x,m=x}^{z,z} S_{ij,lm} R^{\alpha}_{ij,lm}. \tag{4.24}$$

Every second-rank tensor can be decomposed in a sum of three terms respectively a symmetric traceless tensor, an antisymmetric tensor and a scalar the trace:

$$\begin{aligned} S_{ij,lm} &= \left[\frac{1}{2}\left(S_{ij,lm} + S_{ij,ml}\right) - \frac{1}{3}\delta_{lm}\sum_{k=x}^{z} S_{ij,kk}\right] \\ &\quad + \left[\frac{1}{2}\left(S_{ij,lm} - S_{ij,ml}\right)\right] + \frac{1}{3}\delta_{lm}\sum_{k=x}^{z} S_{ij,kk} \end{aligned} \tag{4.25}$$

where $S^{s}_{ij,lm} \equiv \left[\frac{1}{2}\left(S_{ij,lm} + S_{ij,ml}\right) - \frac{1}{3}\delta_{lm}\sum_{k=x}^{z} S_{ij,kk}\right]$ is a symmetric traceless tensor, $S^{a}_{ij,lm} \equiv \left[\frac{1}{2}\left(S_{ij,lm} - S_{ij,ml}\right)\right]$ is an antisymmetric tensor and $S^{tr}_{ij} \equiv \sum_{k=x}^{z} S_{ij,kk}$ is a scalar, the trace of $S_{ij,lm}$. With the same decomposition applied to the tensor $R^{\alpha}_{ij,lm}$, $\mathcal{H}_{1,\alpha}$ becomes:

$$\mathcal{H}_{1,\alpha} = \sum_{i=1,\,j>i}^{N-1,N} \left\{ \sum_{l=x,m=x}^{z,z} \left[S^{s}_{ij,lm} R^{\alpha,s}_{ij,lm} + S^{a}_{ij,lm} R^{\alpha,a}_{ij,lm}\right] + S^{tr}_{ij} R^{\alpha,tr}_{ij} \right\} \tag{4.26}$$

Considering without loss of generality the external magnetic induction $\mathbf{B}_0$ along the Z axis of the laboratory frame, $\mathbf{B}_0 = B_0 \hat{\mathbf{e}}_z$ and introducing the raising and lowering spin operators $I_{k,+} \equiv I_{k,x} + i I_{k,y}$ and $I_{k,-} \equiv I_{k,x} - i I_{k,y}$ it is possible to show that in first-order perturbation theory the different contributions to $\mathcal{H}_1$ become:

$$\begin{aligned} \mathcal{H}_{\sigma} &= \sum_{k=1}^{N} \gamma_k \hbar I_{k,z} \sigma_{k,zz} B_{0,z} = \sum_{k=1}^{N} \gamma_k \hbar I_{k,z} \left(\sigma^{s}_{k,zz} + \frac{1}{3}\sigma^{tr}_{k}\right) B_{0,z} \\ \mathcal{H}_{J} &= 2\pi\hbar \sum_{k=1,\,l>k}^{N-1,N} \left\{ \left[I_{k,z} I_{l,z} - \frac{1}{4}(I_{k,+} I_{l,-} + I_{k,-} I_{l,+})\right] J^{s}_{kl,zz} \right. \\ &\quad \left. + \left[I_{k,z} I_{l,z} + \frac{1}{2}(I_{k,+} I_{l,-} + I_{k,-} I_{l,+})\right] \frac{1}{3} J^{tr}_{kl} \right\} \end{aligned}$$

$$\mathcal{H}_D = \frac{\mu_0}{4\pi} \sum_{k=1,l>k}^{N-1,N} \left\{ \left[I_{k,z} I_{l,z} - \frac{1}{4}(I_{k,+} I_{l,-} + I_{k,-} I_{l,+}) \right] \times \frac{\gamma_k \gamma_l \hbar^2}{r_{kl}^3} \left(1 - \frac{3}{r_{kl}^2} r_{kl,z}^2 \right) \right\}$$

$$\mathcal{H}_Q = \sum_{k=1}^{N} \frac{3eQ_k}{4I_k(2I_k - 1)} \left\{ \left[I_{k,z}^2 - \frac{1}{3} I(I+1) \right] V_{k,zz} \right\} \quad (4.27)$$

4.4.2 Reduced Spin Hamiltonian in High-Field NMR Spectroscopy of Isotropic Liquids and Gases

When the spin system under study belongs to a gas or an isotropic fluid, another very important simplification of the spin Hamiltonian arises. In these cases, the fast molecular motions average to zero the anisotropic interactions and in spectroscopy analysis where the energy levels of the total Hamiltonian are being probed, only the isotropic chemical shift and the isotropic indirect spin–spin interaction remain leading to

$$\mathcal{H}_Z = -\sum_{k=1}^{N} \gamma_k \hbar I_{k,z} B_{0,z}$$

$$\mathcal{H}_\sigma = \sum_{k=1}^{N} \gamma_k \hbar I_{k,z} \frac{1}{3} \sigma_k^{tr} B_{0,z}$$

$$\mathcal{H}_J = 2\pi\hbar \sum_{k=1,l>k}^{N-1,N} \left[I_{k,z} I_{l,z} + \frac{1}{2}(I_{k,+} I_{l,-} + I_{k,-} I_{l,+}) \right] \frac{1}{3} J_{kl}^{tr} \quad (4.28)$$

4.4.3 Selection Rules and Transition Probabilities

In modern NMR spectroscopy analysis the resonance frequencies of the spin system are probed by RF pulse techniques, where an oscillating magnetic induction with a frequency meeting the Bohr condition is applied perpendicular to the static $\mathbf{B}_0$ field for short time periods. This gives rise to a time-dependent perturbation term in the spin Hamiltonian amounting to:

$$\mathcal{H}_2(t) = -\sum_{k=1}^{N} \gamma_k \hbar I_{k,x} B(t)_{1,x} \tag{4.29}$$

where it was considered that $\mathbf{B}_1 = B_1 \mathbf{e}_x$. This term includes the spin operators $I_{k,x}$ that only have nonzero expectation values $< \psi_a | I_{k,x} | \psi_b >$ between states $|\psi_a >$ and $|\psi_b >$ differing by

$$\Delta m_k = \pm 1 \tag{4.30}$$

and consequently only the transitions obeying 4.30 are induced by $\mathcal{H}_2$. Equation 4.30 is the selection rule for nuclear spin transitions induced by a high-frequency induction field meeting the Bohr condition. The transition probability per unit time induced by the high-frequency field may be obtained from time-dependent perturbation theory and is given by (Abragam, 1961)

$$W_{\alpha\beta} = \frac{2\pi}{\hbar} | < \alpha | \mathcal{H}_2 | \beta > |^2 \delta(E_\alpha - E_\beta - \hbar\omega_i) \tag{4.31}$$

where ω_i is the Larmor frequency for the spin i. In practice the delta function in 4.31 becomes replaced by a shape function arising from the finite width of the resonance lines. We have gathered all the ingredients required for a first determination of the NMR spectra associated with a given nuclear spin system and experimentally observable either with a pulsed Fourier transform NMR experiment using a nonselective high-frequency pulse or a CW NMR experiment.

4.4.4 Energy Spectrum of the Nuclear Spin Hamiltonian in High-Field NMR Spectroscopy

The determination of the energy spectrum of the nuclear spin Hamiltonian requires one to find the eigenvalues and eigenvectors of the spin Hamiltonian. In solids this may be an impossible task to carry out by direct methods due to the prohibitive number of interacting spins, and approximation methods become the only option. In soft matter particularly liquids and also gases the molecules undergo rapid rotational and translational diffusion motions; the perturbing Hamiltonian $\mathcal{H}_1 = \mathcal{H}_\sigma + \mathcal{H}_J + \mathcal{H}_D + \mathcal{H}_Q$ observed with an NMR experiment is the result of an average over these fast motions and $\mathcal{H}_1$ is replaced in these systems by $\bar{\mathcal{H}}_1$

(Emsley, 1983). Two terms in $\bar{\mathcal{H}}_1$ include couplings between pairs of spins, $\bar{\mathcal{H}}_J$ and $\bar{\mathcal{H}}_D$, one consequence of the averaging process resulting from the fast motions is that the interactions between two spins belonging to two different molecules are averaged to zero. Each molecule is still a complicated spin system, but $\bar{\mathcal{H}}_J$ and $\bar{\mathcal{H}}_D$ are inversely proportional to r cube, where r is the distance between two interacting spins, this fact enables one to consider only small groups of interacting spins in each molecule. Each molecule will have in general several of these groups of interacting spins with the identical molecules equivalent to each other. The study of the energy spectrum of the Hamiltonian for a N spin system in soft matter systems is performed by studying the Hamiltonian for each one of the n subsystems of interacting spins in each molecule disregarding the interactions between the n subsystems (BOS et al., 1980; HSI et al., 1978). The total Hamiltonian is a sum of the Hamiltonians for each subsystem. The interactions between the n subsystems themselves and the subsystems and the lattice can be neglected in calculating the energy spectrum of the Hamiltonian because they only contribute by broadening the energy levels and hence the resonance lines. On the other hand these neglected interactions are responsible for bringing the spin system to equilibrium with the lattice at a certain temperature T (Abragam, 1961; Slichter, 1992). In the study of relaxation processes, these interactions play a dominant role.

4.4.4.1 Energy spectrum of a system of *n* interacting spins

The Hamiltonian takes the form:

$$\bar{\mathcal{H}} = \mathcal{H}_0 + \bar{\mathcal{H}}_1 \tag{4.32}$$

where $\mathcal{H}_0$ is the Zeeman interaction and $\bar{\mathcal{H}}_1$ is the perturbation Hamiltonian to be dealt with in first-order perturbation theory. Considering the external induction field $\vec{B}_0$ along the z axis of a laboratory fixed frame, the eigenstates of $\mathcal{H}_0$ are tensor product states of eigenstates of $(I_k)^2$ and $I_{k,z}$, $|I_k, m_i>$:

$$|F, m_1, m_2, \ldots >= |I_1, m_1 > |I_1, m_2 > \ldots |I_n, m_n > \tag{4.33}$$

Each eigenstate $|I_k, m_k>$ obeys the equations

$$
\begin{aligned}
I_{k,z}|I_k, m_k> &= m_k|I_k, m_k> \\
I_k^2|I_k, m_k> &= I_k\,(I_k+1)\,|I_k, m_k> \\
I_{k,+}|I_k, m_k> &= (I_{k,x} + i\,I_{k,y})|I_k, m_k> \\
&= \sqrt{I_k(I_k+1) - m_k(m_k+1)}|I_k, m_k+1> \\
I_{k,-}|I_k, m_k> &= (I_{k,x} - i\,I_{k,y})|I_k, m_k> \\
&= \sqrt{I_k(I_k+1) - m_k(m_k-1)}|I_k, m_k-1>
\end{aligned}
\quad (4.34)
$$

and consequently the eigenvalue equation for $\mathcal{H}_0$ becomes:

$$
\begin{aligned}
\mathcal{H}_0|F, m_1, m_2, \ldots> &= -\hbar B_{0,z}(\gamma_1 m_1 + \gamma_2 m_2 + \ldots)|F, m_1, m_2, \ldots> \\
E_{0,m_1,m_2,\ldots} &= -\hbar B_{0,z}(\gamma_1 m_1 + \gamma_2 m_2 + \ldots).
\end{aligned}
\quad (4.35)
$$

When the spin system is composed by only one spin species, γ_i is the same for all spins and can be factorized in Eq. 4.35. In this case the spacing between two adjacent energy levels of $\mathcal{H}_0$ corresponding to the change of just one unit in the magnetic quantum number of one of the spins is $\hbar\gamma B_{0,z}$ which corresponds to a photon frequency of $w = \gamma B_{0,z}$. This is the Larmor precession frequency for that spin species in the field of intensity $B_{0,z}$. The energy absorption spectrum of $\mathcal{H}_0 + \bar{\mathcal{H}}_1$ in this case is composed of absorption lines centered around w, since $\bar{\mathcal{H}}_1$ shifts the energy levels of $\mathcal{H}_0$ but by an amount much smaller than their spacing in the absence of $\bar{\mathcal{H}}_1$. When more than one nuclear spin species is present the energy levels of $\bar{\mathcal{H}}_0$ can be expressed as:

$$
E^0_{m_{1,1},m_{1,2},\ldots m_{L,p_L}} = -\sum_{i=1}^{L} \hbar B_{0,z}\gamma_i \sum_{j=1}^{p_i} m_{i,j} \quad (4.36)
$$

where L is the number of different nuclear spin species present in the system and p_i is the number of spins of the ith species. In this case a different Larmor precession frequency is defined for each spin species $w_i = \gamma_i B_{0,z}$ since a transition of a spin of the ith species between states with magnetic quantum numbers m_i and $m_i + 1$ produces a photon of energy $\hbar w_i = \hbar\gamma_i B_{0,z}$. The energy absorption spectrum of $\bar{\mathcal{H}}$ is composed of several subspectra, one for each spin species, each one of them centered at the Larmor precession frequency for that spin species.

The first-order corrections to E_0 introduced by $\bar{\mathcal{H}}_1$ for nondegenerate states are given by:

$$E^1_{m_1,m_2,\ldots} =< F, m_1, m_2, \ldots |\bar{\mathcal{H}}_1| F, m_1, m_2, \ldots > \qquad (4.37)$$

For degenerate cases, the standard approach of diagonalizing $\bar{\mathcal{H}}_1$ inside the subspaces with the same E_0 is used. The eigenvalues thus found are the first-order energy corrections to E_0 associated with corresponding eigenstates. The use of the method is exemplified in the next chapter.

4.5 Dynamics of Interacting Spins

4.5.1 The Density Matrix

When addressing the dynamics of interacting spins one must go beyond Bloch's equations as the determination of the time evolution of the NMR observables requires a quantomechanical analysis. A central tool in this process is the density matrix as it allows the determination of the expectation values of the sought NMR observables. Considering a dynamical variable represented by the operator A and the system's wave function $|\psi(t) >= \sum_{i=1}^{n} C_i(t)|u_i >$ where the kets $|u_i >$ are the eigenstates of the spin Hamiltonian and the set of time-dependent complex numbers $C_i(t)$ represents $|\psi(t) >$ in the $\{|u_i >\}$ basis, one obtains the expectation value for A in the state defined by $|\psi(t) >$ as

$$< \psi|A|\psi >= \sum_{i=1,j=1}^{n,n} C_i^*(t)C_j(t) < u_i|A|u_j > . \qquad (4.38)$$

The product of coefficients $C_i^*(t)C_j(t)$ appearing in 4.38 can be considered as the matrix element of the operator $\rho(t) = |\psi(t) >< \psi(t)|$ which is known as the density operator (Cohen-Tannoudji et al., 1977). Its matrix representation in the $\{|u_i >\}$ basis, $[\rho(t)]$ whose elements are

$$\rho_{ji}(t) =< u_j|\psi(t) >< \psi(t)|u_i >= C_i^*(t)C_j(t), \qquad (4.39)$$

allows the expectation value $< \psi|A|\psi >$ to be written as:

$$\begin{aligned} < \psi|A|\psi > &= \sum_{i=1,j=1}^{n,n} < u_i|A|u_j > \rho_{ji} = \sum_{i=1,j=1}^{n,n} A_{ij}\rho_{ji} \\ &= \mathrm{Tr}\{A\rho\} \end{aligned} \qquad (4.40)$$

Equation 4.40 shows that in a system described by the wave function $|\psi(t) >$ the expectation value of A can be calculated evaluating the trace of the product of A by the density matrix $\rho(t)$. In many real cases a complete information about the state of the system is not available and a probabilistic approach must be considered. In these cases the system is described by a statistical mixture of states $|\psi_1 >, |\psi_2 >, \ldots, |\psi_N >$ with probabilities $p_1, p_2, \ldots, p_N$ with $\sum_{i=1}^{N} p_i = 1$. This is precisely what happens in a system in thermodynamic equilibrium at a temperature T, which has a probability proportional to $e^{-E_i/kT}$ of being in a state with energy E_i. The density matrix for a system described by a statistical mixture of states can be obtained from the evaluation of the expectation value of the operator A in similar terms as was done in 4.40 for the pure case. For a statistical mixture of states, the expectation value of the observable A is given by

$$\begin{aligned} \langle A \rangle &= \sum_{i=1}^{N} p_i < \psi_i|A|\psi_i >= \sum_{i=1}^{N} \sum_{j=1,k=1}^{n,n} A_{jk} \left(p_i \rho_{kj,i} \right) \\ &= \sum_{j=1,k=1}^{n,n} A_{jk} \left(\sum_{i=1}^{N} p_i \rho_{kj,i} \right) = Tr\{A\bar{\rho}\} \end{aligned} \quad (4.41)$$

with $\bar{\rho}$ given by

$$\bar{\rho_{jk}}(t) = \sum_{i=1}^{N} p_i \rho_{jk,i} = \sum_{i=1}^{N} p_i C_k^{i*}(t) C_j^i(t) = \overline{C_k^*(t) C_j(t)}. \quad (4.42)$$

Owing to its definition, Eq. 4.42 for a statistical ensemble or Eq. 4.39 for a pure case, the density matrix is a Hermitian operator and consequently has real eigenvalues. Its diagonal elements $\bar{\rho_{ii}} = \overline{C_i^*(t) C_i(t)}$ are positive real numbers and represent the populations of states $|u_i >$ in the ensemble. The off diagonal elements $\bar{\rho_{ij}}$,with $i \neq j$ represent coherences between the states $|u_i >$ and $|u_j >$.

4.5.1.1 Time evolution of the density matrix: the Liouville/von Neumann equation

The determination of the expectation value of an observable A of the spin system at time t given by the trace formula $\langle A \rangle = Tr\{A\bar{\rho}\}$, requires the knowledge of the density matrix at that time $\bar{\rho}(t)$. This can be achieved solving the Liouville/von Neumann equation which determines the time evolution of the density matrix (Ernst et al., 1992; Kimmich, 1997)

$$i\hbar\frac{d\bar{\rho}(t)}{dt} = [\mathcal{H}(t), \bar{\rho}(t)]\,. \tag{4.43}$$

Equation 4.43 which is also valid for a pure case, follows directly from the Schrodinger equation $i\hbar\frac{d|\psi(t)>}{dt} = \mathcal{H}(t)|\psi(t) >$ and it is very important in modern pulsed NMR. This equation conveniently describes the evolution of the spin system subjected to different NMR pulse sequences which are at the center of NMR spectroscopy and relaxometry. When the Hamiltonian is time independent, Eq. 4.43 has a solution given by

$$\bar{\rho}(t) = e^{\{\frac{-i}{\hbar}\mathcal{H}t\}}\bar{\rho}(0)e^{\{\frac{i}{\hbar}\mathcal{H}t\}}. \tag{4.44}$$

The matrix representation of $\bar{\rho}(t)$ in the basis $\{|u_i >\}$ of eigenvectors of the Hamiltonian $\mathcal{H}$, considered here to be time independent, such that

$$\mathcal{H}|u_i >= E_i|u_i >, \tag{4.45}$$

may be found from Eq. 4.44, leading to

$$\bar{\rho_{ii}}(t) = \bar{\rho_{ii}}(0)\bar{\rho_{ij}}(t) = \bar{\rho_{ij}}(0)e^{\{\frac{-i}{\hbar}(E_i-E_j)t\}}. \tag{4.46}$$

This shows that the diagonal elements are time independent, while the off-diagonal elements oscillate with a frequency corresponding to the energy difference between the two states involved.

4.5.1.2 The density matrix in thermodynamic equilibrium

According to quantum statistical mechanics (Huang, 1987) the density matrix for a system in thermal equilibrium takes the value

$$\bar{\rho} = Z^{-1}e^{-\mathcal{H}/kT} \tag{4.47}$$

where $\mathcal{H}$ is the system Hamiltonian, k is the Boltzmann constant and Z is a normalization coefficient required for the trace of $\bar{\rho}$ to be 1. Z

is called the partition function

$$Z = Tr\{e^{-\mathcal{H}/kT}\}. \tag{4.48}$$

In the basis $\{|u_i>\}$ of eigenvectors of the Hamiltonian $\mathcal{H}$, considered here time independent, the elements of the density matrix take the values

$$\begin{aligned} \bar{\rho}_{ii} &= Z^{-1} < u_i|e^{-\mathcal{H}/kT}|u_i >= Z^{-1}e^{-E_i/kT} \\ \bar{\rho}_{ij} &= Z^{-1} < u_i|e^{-\mathcal{H}/kT}|u_j >= Z^{-1}e^{-E_j/kT} < u_i|u_j >= 0. \end{aligned} \tag{4.49}$$

4.6 Pulsed Nuclear Magnetic Resonance

Modern NMR relies on pulsed experiments to investigate the spin system (Ernst et al., 1992; Kimmich, 1997; Slichter, 1992). The method was fostered by the theoretical result of Lowe and Norberg (Lowe and Norberg, 1957) showing that the Fourier transform of the induction signal from the transverse magnetization decay (usually referred to as *free induction decay*, FID) following a $\pi/2$ pulse in solids with dipolar broadening produced the absorption spectra of the spin system, and the work of Ernst and Anderson (Ernst and Anderson, 1966) showing that the use of RF pulses has significant experimental advantages over steady-state methods.

4.6.1 Detection of the NMR Observable

In pulsed NMR with quadrature detection the quantity measured is proportional to the complex transverse nuclear magnetization. When the external static magnetic induction B_0 is aligned along z the transverse magnetization arising from a particular spin species takes the form

$$M_x + iM_y = N\gamma\hbar\left\langle I_x + iI_y\right\rangle = N\gamma\hbar\left\langle I_+\right\rangle = N\gamma\hbar Tr\{\bar{\rho}I_+\}, \tag{4.50}$$

where N is the number of targeted spins per unit volume, γ is the gyromagnetic ratio, and $\bar{\rho}(t)$ is the density matrix. The transverse nuclear magnetization signal is recorded by an RF-picking coil with its axis perpendicular to the static B_0 field and located around the NMR sample. At thermodynamic equilibrium the population of the stationary states given by the diagonal elements of $\bar{\rho}$ are decreasing exponential functions of the states' energy, while the coherences between the states given by the off-diagonal elements are zero. Under these conditions the trace operation appearing in 4.50 produces a null result and no NMR signal is detected. To observe the transverse complex magnetization it is necessary to drive the spin system out of thermodynamic equilibrium and this is accomplished by the application of appropriate RF pulses of specific duration and amplitude which are generated by energizing the same RF coil used in detection prior to the detection process.

4.6.2 Time Evolution of the Density Matrix under RF Pulses

In pulsed NMR experiments the time evolution of the nuclear spins system is conditioned by the spin interactions and the application of appropriate RF pulses, these pulses are responsible for inducing changes in the populations of the energy levels of the spin system and the creation of coherences between the states. The general equation of motion for the density matrix is the Liouville/von Neumann Eq. 4.43 which has the solution 4.44 when the Hamiltonian is time independent. As seen when studying the motion of noninteracting spins 4.3, it is convenient to analyze the evolution of the spin system in the rotating frame introduced in 4.3.1. In this frame $\{x', y', z'\}$ with its z' axis coinciding with the z axis of the laboratory fixed frame but rotating around it with a uniform angular velocity Ω, which at resonance takes the value $\Omega = -\gamma B_0 \mathbf{e}_z$, the Hamiltonian becomes simplified and when relaxation processes can be disregarded in short time lapses the Hamiltonian may become time independent in those lapses allowing 4.44 as a valid solution. The transformation to the rotating frame of the relevant operators is carried out through the unitary operator

$$R \equiv e^{-i\omega t I_z} \tag{4.51}$$

with ω the rotation frequency and I_z the z component of the total spin operator. The density matrix in the rotating frame is

$$\bar{\rho}'(t) = e^{-i\omega t I_z} \bar{\rho}(t) e^{i\omega t I_z} \tag{4.52}$$

and the Liuoville/von Neumann equation becomes

$$i\hbar\frac{d\bar{\rho}'(t)}{dt} = \left[\mathcal{H}_e, \bar{\rho}'(t)\right], \tag{4.53}$$

where

$$\mathcal{H}_e = R\mathcal{H}R^{-1} + \omega\hbar I_z. \tag{4.54}$$

To evaluate $\mathcal{H}_e$ we note that the system Hamiltonian $\mathcal{H}$ is composed of three contributions which are, respectively, the Zeeman term $\mathcal{H}_0$, the RF term $\mathcal{H}_{\mathrm{RF}}$ and the spin interactions term $\mathcal{H}_i$, leading to

$$\begin{aligned} R\mathcal{H}_0 R^{-1} &= -\gamma\hbar I_z B_0 \\ R\mathcal{H}_{\mathrm{RF}} R^{-1} &= R[-2\gamma\hbar I_x B_1 \cos(\omega t)] R^{-1} \\ &= -\gamma\hbar B_1 I_x - \gamma\hbar B_1 \left[\cos(2\omega t) I_x + sin(2\omega t) I_y\right] \\ R\mathcal{H}_i R^{-1} &= \mathcal{H}'_i. \end{aligned} \tag{4.55}$$

The terms in $R\mathcal{H}_{\mathrm{RF}} R^{-1}$ oscillating with frequency 2ω are not effective in inducing magnetization changes and can be neglected. At resonance, where $\omega = \gamma B_0$, the term arising from $R\mathcal{H}_0 R^{-1}$, is canceled by the term $\omega\hbar I_z$ in Eq. 4.54, giving for the effective Hamiltonian the form

$$\mathcal{H}_e = -\omega_1\hbar I_x + \mathcal{H}'_i \tag{4.56}$$

with $\omega_1 = \gamma B_1$. $\mathcal{H}_e$ given by Eq. 4.56 is composed of two terms, one originated by the RF pulses $-\omega_1\hbar I_x$ and the other arising from the spin interactions $\mathcal{H}'_i$. The time dependence of $\mathcal{H}_e$ comes from the term $\mathcal{H}'_i$ and arises due to the modulation of spin interactions created by the molecular motions and reorientations. When the RF pulses are hard pulses, the case discussed here, and while an RF pulse is being applied the term $-\omega_1\hbar I_x$ dominates $\mathcal{H}_e$ and the spin interactions term can be neglected. During this period a solution of the type Eq. 4.44 for Eq. 4.53 applies. In absence of RF pulses $\mathcal{H}'_i$ which is time dependent in general can for certain time lapses be replaced by an averaged Hamiltonian enabling also a solution of the type Eq. 4.44 to be applicable (Kimmich, 1997). Under these circumstances where $\mathcal{H}_e$ takes different time-independent forms during different time intervals $\tau_1, \tau_2 \ldots t$, the solution of Eq. 4.53 can be written as

$$\begin{aligned} &\bar{\rho}'(t + \ldots + \tau_2 + \tau_1) = e^{\{\frac{-i}{\hbar}\mathcal{H}_e t\}} \ldots e^{\{\frac{-i}{\hbar}\mathcal{H}_e \tau_2\}} e^{\{\frac{-i}{\hbar}\mathcal{H}_e \tau_1\}} \\ &\bar{\rho}'(0) e^{\{\frac{i}{\hbar}\mathcal{H}_e \tau_1\}} e^{\{\frac{i}{\hbar}\mathcal{H}_e \tau_2\}} \ldots e^{\{\frac{i}{\hbar}\mathcal{H}_e t\}}. \end{aligned} \tag{4.57}$$

An example of this method appears in the next chapter. When $\mathcal{H}_e$ is time dependent and cannot be substituted in successive time intervals by time-independent approximations, specific solution methods for Eq. 4.53 with a time-dependent $\mathcal{H}_e$ must be used, which fall outside the scope of this introduction, and the reader is referred to more specialized literature (Ernst et al., 1992; Kimmich, 1997; Slichter, 1992).

Chapter 5

NMR Spectroscopy of Anisotropic Fluid Systems: Theory and Experiment

5.1 Introduction

In the study of anisotropic fluids using high-field nuclear magnetic resonance (NMR) spectroscopy several nuclei carrying a spin can be used as probes. The most used probes in anisotropic fluids are the nuclei of hydrogen, deuterium, carbon 13, nitrogen, and fluorine. Hydrogen with a nucleus of spin 1/2 has the advantage of giving high signal-to-noise ratios because of its high gyromagnetic ratio and also because it is present in high numbers in the system's molecules. It has the disadvantage that the information obtained is nonselective because many different hydrogens in a molecule contribute to the response of the system. Also the high number of hydrogen nuclei interacting through dipolar coupling significantly hampers a detailed simulation of the data. Deuterium with a nucleus of spin 1 is not present naturally in anisotropic fluid molecules, and its use requires the replacement of hydrogen atoms in the molecule by deuterium atoms, a process called deuteration. The small number

NMR of Liquid Crystal Dendrimers
Carlos R. Cruz, João L. Figueirinhas, and Pedro J. Sebastião

ISBN 978-981-4745-72-7 (Hardcover), 978-981-4745-73-4 (eBook)
www.panstanford.com

of interacting particles necessary to consider when deuterium is the probe nucleus enables a full and unique interpretation of the spectroscopic data. Nitrogen and fluorine are less used; they are present in much smaller quantities in anisotropic fluid molecules or not present at all. For some of these nuclei short spin–lattice relaxation times present in these probes broaden the spectral lines, making the interpretation of the data difficult.

5.2 Nuclear Spin Hamiltonian for NMR of Anisotropic Fluid Systems

The relevant nuclear spin Hamiltonian for NMR of anisotropic fluid systems includes beyond the Zeeman term $\mathcal{H}_z$ the four terms introduced in the previous chapter, Section 4.4; these terms are the chemical shift $\mathcal{H}_\sigma$, the indirect spin–spin coupling $\mathcal{H}_J$, the direct dipolar interaction $\mathcal{H}_D$, and the quadrupolar interaction $\mathcal{H}_Q$, present for spins with $I \geq 1$. As discussed in the previous chapter, Section 4.4.1, in high-field NMR spectroscopy, the energy levels of the nuclear spin Hamiltonian can be determined in first-order perturbation theory, considering the Zeeman term $\mathcal{H}_z$ as the unperturbed Hamiltonian $\mathcal{H}_0 = \mathcal{H}_z$ and the remaining terms as the perturbation $\mathcal{H}_1 = \mathcal{H}_\sigma + \mathcal{H}_J + \mathcal{H}_D + \mathcal{H}_Q$. Due to the rapid molecular reorientation and diffusional motions occurring in anisotropic fluid systems the contribution of $\mathcal{H}_1$ for the system's Hamiltonian is actually replaced by a time average of this quantity $\bar{\mathcal{H}}_1$. As discussed in Section 4.4.4 the molecular motions decouple the spins from different molecules in terms of energy levels of the Hamiltonian and due to the $1/r^3$ dependence of the Hamiltonian terms involving the interaction of two spins r apart, it is frequently possible to separate the spins of a molecule in several groups of few interacting spins each and analyze each group separately. The total Hamiltonian is the sum of the Hamiltonians from each spin group in the molecules with the different molecules equivalent to each other. From Eq. 4.27 the perturbing Hamiltonian terms arising from a group of n spins is

$$\bar{\mathcal{H}}_\sigma = \sum_{k=1}^{n} \gamma_k \hbar I_{k,z} \bar{\sigma}_{k,zz} B_0 = \sum_{k=1}^{n} \gamma_k \hbar I_{k,z} \left(\bar{\sigma^s}_{k,zz} + \frac{1}{3} \sigma_k^{tr} \right) B_0$$

$$\bar{\mathcal{H}}_J = 2\pi\hbar \sum_{k=1,l>k}^{n-1,n} \left\{ \left[I_{k,z}I_{l,z} - \frac{1}{4}(I_{k,+}I_{l,-} + I_{k,-}I_{l,+}) \right] \bar{J}^{s}_{kl,zz} + \left[I_{k,z}I_{l,z} + \frac{1}{2}(I_{k,+}I_{l,-} + I_{k,-}I_{l,+}) \right] \frac{1}{3} J^{tr}_{kl} \right\}$$

$$\bar{\mathcal{H}}_D = \frac{\mu_0}{4\pi} \sum_{k=1,l>k}^{n-1,n} \left\{ \left[I_{k,z}I_{l,z} - \frac{1}{4}(I_{k,+}I_{l,-} + I_{k,-}I_{l,+}) \right] \overline{\frac{\gamma_k \gamma_l \hbar^2}{r^3_{kl}} \left(1 - \frac{3}{r^2_{kl}} r^2_{kl,z} \right)} \right\}$$

$$\bar{\mathcal{H}}_Q = \sum_{k=1}^{n} \frac{3eQ_k}{4I_k(2I_k - 1)} \left\{ \left[I^2_{k,z} - \frac{1}{3} I(I+1) \right] \bar{V}_{k,zz} \right\} \tag{5.1}$$

Al these terms have a similar form, they are constituted by the product of spin operators by the zz component or the trace of phenomenological second rank tensors related to system properties, these tensor components are

$$\bar{\sigma}^{s}{}_{k,zz}, \sigma^{tr}_{k},$$
$$\bar{J}^{s}_{kl,zz}, J^{tr}_{kl},$$
$$\overline{\frac{\gamma_k \gamma_l}{r^3_{kl}} \left(1 - \frac{3}{r^2_{kl}} r^2_{kl,z} \right)},$$
$$\bar{V}_{k,zz}. \tag{5.2}$$

The time averages of the zz components of the different second rank tensorial quantities associated with the specific molecular properties contain information on molecular structure and orientational order and can be given in terms of the principal values of the corresponding second rank tensorial quantities as follows

$$\bar{V}_{k,zz} = \bar{V}_{k,x'x'} \sin^2(\theta)\cos^2(\phi) + \bar{V}_{k,y'y'} \sin^2(\theta)\sin^2(\phi) + \bar{V}_{k,z'z'} \cos^2(\theta). \tag{5.3}$$

The quantities $\bar{V}_{k,x'x'}$, $\bar{V}_{k,y'y'}$, $\bar{V}_{k,z'z'}$ are the principal values of the tensor $\bar{V}_k$, the primed frame $[x', y', z']$ is the principal frame of that tensor and the angles θ and ϕ are the polar and azimuthal angles defining the orientation of the B_0 field in the principal frame. Introducing the asymmetry parameter η for the tensor $\bar{V}_k$,

$$\eta = \frac{\bar{V}_{k,x'x'} - \bar{V}_{k,y'y'}}{\bar{V}_{k,z'z'}} \tag{5.4}$$

and considering that the tensor $\bar{V}_k$ is traceless, $\bar{V}_{k,zz}$ can be written in a more compact form as

$$\bar{V}_{k,zz} = \bar{V}_{k,z'z'} \left[\frac{3}{2} \cos^2(\theta) - \frac{1}{2} + \frac{\eta}{2} \sin^2(\theta) \cos(2\phi) \right]. \quad (5.5)$$

A corresponding development applies to the other tensorial quantities in Eq. 5.1. All those quantities can also be related to a set of orientational order parameters characterizing the anisotropic fluid.

The two terms σ_k^{tr} and J_{kl}^{tr} are scalars and carry information on the molecular structure, they are the only nonzero terms in isotropic systems and are discussed in 4.4.2. The quantities listed in Eq. 5.2 can in principle be obtained from the energy spectrum of the spin Hamiltonian.

5.3 Averaged Second-Rank Tensorial Quantities and Order Parameters

In order to determine the order parameters involved in the relationship between the components of an averaged second rank tensorial quantity in a lab. fixed frame and the components of the tensorial quantity in the molecular fixed frame, we start by considering a traceless and symmetric second rank tensorial quantity $R^M_{\alpha\beta}$ in a frame "M" fixed with a rigid molecular segment. We then use the rules of tensor calculus to establish the relation between the components of the tensorial quantity in the "M" frame and in a frame "F," named phase frame, fixed with the external magnetic induction of the NMR spectrometer and take a time average of the result as the orientation of the "M" frame is changing in time relative to the "F" frame due to molecular motions leading to

$$\bar{R}^F_{ij} = \sum_{\alpha\beta} \overline{\frac{\partial x_\alpha}{\partial x_i} \frac{\partial x_\beta}{\partial x_j} R^M_{\alpha\beta}}. \quad (5.6)$$

Considering the direction cosines defining the orientation of frame "M" in frame "F," the quantities $\frac{\partial x_\alpha}{\partial x_i} \frac{\partial x_\beta}{\partial x_j}$ become

$$\frac{\partial x_\alpha}{\partial x_i} \frac{\partial x_\beta}{\partial x_j} = \cos(\theta_{i\alpha}) \cos(\theta_{j\beta}), \quad (5.7)$$

leading to

$$\bar{R}_{ij}^{F} = \sum_{\alpha\beta} \overline{\cos(\theta_{i\alpha})\cos(\theta_{j\beta})R_{\alpha\beta}^{M}}. \tag{5.8}$$

The order matrix defined by

$$S_{\alpha\beta}^{ij} = \frac{1}{2} < 3\cos(\theta_{i\alpha})\cos(\theta_{j\beta}) - \delta_{\alpha\beta}\delta_{ij} > \tag{5.9}$$

(de Gennes and Prost, 1993) where the brackets <> indicate an ensemble average, can be used in certain cases to express $\bar{R}_{ij}^{F}$ (Eq. 5.8) in terms of the order matrix elements. One of these cases corresponds to a constant $R_{\alpha\beta}^{M}$ tensor, in this case, inverting Eq. 5.9 leads to

$$< \cos(\theta_{i\alpha})\cos(\theta_{j\beta}) >= \frac{1}{3}(2S_{\alpha\beta}^{ij} + \delta_{\alpha\beta}\delta_{ij}) \tag{5.10}$$

and using the ergodic properties of the system the ensemble average replaces the time average in $\bar{R}_{ij}^{F}$ and Eq. 5.10 is used, leading to

$$\bar{R}_{ij}^{F} = \frac{2}{3}\sum_{\alpha\beta} S_{\alpha\beta}^{ij} R_{\alpha\beta}^{M}. \tag{5.11}$$

Another case where $\bar{R}_{ij}^{F}$ can be expressed in terms of the order matrix elements corresponds to the situation where $R_{\alpha\beta}^{M}$ tensor is time dependent due to the presence of molecular conformation changes. Assigning a probability p_k and an order matrix $S_{\alpha\beta}^{k,ij}$ to each conformer leads to

$$\bar{R}_{ij}^{F} = \frac{2}{3}\sum_{\alpha\beta}\sum_{k} p_k S_{\alpha\beta}^{k,ij} R_{\alpha\beta}^{k,M}. \tag{5.12}$$

The order matrix $S_{\alpha\beta}^{ij}$ is a symmetric traceless tensor in indices ij and $\alpha\beta$, from its 91 elements only 34 at most are independent (de Gennes and Prost, 1993). The number of independent elements of the order matrix is further reduced by the symmetry of the molecules and the symmetry of the anisotropic fluid phase. When determining order matrix elements using any of the NMR anisotropic observables listed in Eq. 5.2 in the case of constant $R_{\alpha\beta}^{M}$, it is necessary to choose both the phase frame "F" and the molecular frame "M." As the number of NMR observables may be limited, it is convenient to choose for "F" the frame that diagonalizes $S_{\alpha\beta}^{ij}$ in the lab. related indices ij and choose for "M" the frame that diagonalizes

$S^{ii}_{\alpha\beta}$ in the molecule related indices $\alpha\beta$, with i one of the principal directions of $S^{ij}_{\alpha\beta}$ in "F." The symmetry axes of the anisotropic fluid phase are good candidates for the principal directions in "F," while the symmetry axis of the molecular segment may also be good candidates for the principal directions of $S^{ii}_{\alpha\beta}$ in "M." The following example illustrates the method; consider that the principal z axis of $S^{ij}_{\alpha\beta}$ in "F" is along the external magnetic induction field **B** and that through the analysis of molecular symmetry we have identified the principal "M" frame leading to $S^{zz}_{ij} = 0$ $(i \neq j)$ in that frame. Then it is possible to calculate the order matrix elements S^{zz}_{zz} and $S^{zz}_{xx} - S^{zz}_{yy}$ if two independent values for the NMR observable $\bar{R}^{F}_{k,zz}$ are determined from the spectral analysis as the following system of equations shows

$$\begin{aligned}\bar{R}^{F}_{1,zz} &= S^{zz}_{zz} R^{M}_{1,zz} + \frac{1}{3}(S^{zz}_{xx} - S^{zz}_{yy})(R^{M}_{1,xx} - R^{M}_{1,yy}) \\ \bar{R}^{F}_{2,zz} &= S^{zz}_{zz} R^{M}_{2,zz} + \frac{1}{3}(S^{zz}_{xx} - S^{zz}_{yy})(R^{M}_{2,xx} - R^{M}_{2,yy})\end{aligned} \tag{5.13}$$

leading to

$$\begin{aligned}S^{zz}_{zz} &= \frac{\bar{R}^{F}_{1,zz}(R^{M}_{2,xx} - R^{M}_{2,yy}) - \bar{R}^{F}_{2,zz}(R^{M}_{1,xx} - R^{M}_{1,yy})}{\bar{R}^{M}_{1,zz}(R^{M}_{2,xx} - R^{M}_{2,yy}) - \bar{R}^{M}_{2,zz}(R^{M}_{1,xx} - R^{M}_{1,yy})} \\ S^{zz}_{xx} - S^{zz}_{yy} &= 3\frac{\bar{R}^{F}_{2,zz} R^{M}_{1,zz} - \bar{R}^{F}_{1,zz} R^{M}_{2,zz}}{\bar{R}^{M}_{1,zz}(R^{M}_{2,xx} - R^{M}_{2,yy}) - \bar{R}^{M}_{2,zz}(R^{M}_{1,xx} - R^{M}_{1,yy})} \\ S^{zz}_{ij} &= 0, \ (i \neq j)\end{aligned} \tag{5.14}$$

5.4 Determination of High-Field NMR Absorption Spectra for Selected Spin Systems

5.4.1 NMR Spectrum from Two Interacting Identical Spin 1/2 Particles Subjected to the Same Chemical Shift

In spite of being a very simple interacting system, the two interacting spin 1/2 particles has some interest because quite often the hydrogen spectra of some anisotropic fluids can be interpreted as the superposition of spectra arising from several distinct hydrogen pairs in each molecule. Following the method outlined in Chapter 4, Section 4.4.4.1, the eigenstates of the unperturbed Hamiltonian

$\mathcal{H}_0 = \mathcal{H}_z$ for this system are tensor product states of the eigenstates $|I_k, m_i>$ of the operators $(\mathbf{I}_k)^2$ and $\mathbf{I}_{k,z}$ with $k = 1, 2$. The Hilbert space for this system has dimension 4 and the four base states of the form $|I_1, m_1> |I_2, m_2>$ with $m_1 = \pm\frac{1}{2}$ and $m_2 = \pm\frac{1}{2}$ which are eigenstates of the unperturbed Hamiltonian are explicitly

$$
\begin{aligned}
&|\frac{1}{2}, +\frac{1}{2} > |\frac{1}{2}, +\frac{1}{2} > \\
&|\frac{1}{2}, +\frac{1}{2} > |\frac{1}{2}, -\frac{1}{2} > \\
&|\frac{1}{2}, -\frac{1}{2} > |\frac{1}{2}, +\frac{1}{2} > \\
&|\frac{1}{2}, -\frac{1}{2} > |\frac{1}{2}, -\frac{1}{2} >
\end{aligned}
\tag{5.15}
$$

and will be abbreviate to

$$
|++>, |+->, |-+>, |-->
\tag{5.16}
$$

by keeping only the signs of the magnetic quantum numbers of spin 1 and spin 2 in this order. Recalling the relations obeyed by the eigenstates $|I_k, m_i>$, the unperturbed energy levels associated with each state are

$$
\begin{aligned}
E^0_+ &= <++|\mathcal{H}_0|++> = -\gamma\hbar B_0 <++|(I_{1,z} + I_{2,z})|++> \\
&= -\gamma\hbar B_0(<++|I_{1,z}|++> + <++|I_{2,z}|++>) \\
&= -\gamma\hbar B_0 \left(\frac{1}{2} + \frac{1}{2}\right) = -\gamma B_0 \hbar \\
E^0_{0a} &= <+-|\mathcal{H}_0|+-> = -\gamma\hbar B_0 <+-|(I_{1,z} + I_{2,z})|+-> \\
&= -\gamma\hbar B_0 \left(\frac{1}{2} - \frac{1}{2}\right) = 0 \\
E^0_{0b} &= <-+|\mathcal{H}_0|-+> = -\gamma\hbar B_0 <-+|(I_{1,z} + I_{2,z})|-+> \\
&= -\gamma\hbar B_0 \left(-\frac{1}{2} + \frac{1}{2}\right) = 0 \\
E^0_- &= <--|\mathcal{H}_0|--> = -\gamma\hbar B_0 <--|(I_{1,z} + I_{2,z})|--> \\
&= -\gamma\hbar B_0 \left(-\frac{1}{2} - \frac{1}{2}\right) = \gamma\hbar B_0.
\end{aligned}
\tag{5.17}
$$

The states $|++>$ and $|-->$ are nondegenerate, while the states $|+->$ and $|-+>$ are degenerate with $E^0 = 0$. For the nondegenerate states the energy corrections arising from the perturbing Hamiltonian $\bar{\mathcal{H}}_1$ are

$$\begin{aligned} E_+^1 &= <++|\bar{\mathcal{H}}_1|++>=<++|(\bar{\mathcal{H}}_\sigma+\bar{\mathcal{H}}_J+\bar{\mathcal{H}}_D)|++> \\ E_-^1 &= <--|\bar{\mathcal{H}}_1|-->=<--|(\bar{\mathcal{H}}_\sigma+\bar{\mathcal{H}}_J+\bar{\mathcal{H}}_D)|--> \end{aligned} \tag{5.18}$$

where the individual terms give the contributions

$$\begin{aligned} <++|\bar{\mathcal{H}}_\sigma|++> &= \gamma\hbar B_0\bar{\sigma}_{zz} <++|(I_{1,z}+I_{2,z})|++> \\ &= \gamma\hbar B_0\bar{\sigma}_{zz} \\ <--|\bar{\mathcal{H}}_\sigma|--> &= \gamma\hbar B_0\bar{\sigma}_{zz} <--|(I_{1,z}+I_{2,z})|--> \\ &= -\gamma\hbar B_0\bar{\sigma}_{zz} \\ <++|\bar{\mathcal{H}}_J|++> &= 2\pi\hbar[\left(\bar{J}^{s}_{12,zz}+\frac{1}{3}J^{tr}_{12}\right) <++|I_{1,z}I_{2,z}|++> \\ &\quad +\left(-\frac{1}{4}\bar{J}^{s}_{12,zz}+\frac{1}{6}J^{tr}_{12}\right) \\ &\quad <++|(I_{1,+}I_{2,-}+I_{1,-}I_{2,+})|++>] \\ &= 2\pi\hbar\left(\bar{J}^{s}_{12,zz}+\frac{1}{3}J^{tr}_{12}\right)\left(\frac{1}{2}\right)^2 \\ <--|\bar{\mathcal{H}}_J|--> &= 2\pi\hbar\left(\bar{J}^{s}_{12,zz}+\frac{1}{3}J^{tr}_{12}\right)\left(\frac{1}{2}\right)^2 \\ <++|\bar{\mathcal{H}}_D|++> &= \frac{\mu_0}{4\pi}\gamma^2\hbar^2\overline{\frac{1}{r^3_{12}}\left(1-3\frac{r^2_{12,z}}{r^2_{12}}\right)}[<++|I_{1,z}I_{2,z}|++> \\ &\quad -1/4 <++|(I_{1,+}I_{2,-}+I_{1,-}I_{2,+})|++>] \\ &= \frac{\mu_0}{4\pi}\gamma^2\hbar^2\overline{\frac{1}{r^3_{12}}\left(1-3\frac{r^2_{12,z}}{r^2_{12}}\right)}\left(\frac{1}{2}\right)^2 \\ <--|\bar{\mathcal{H}}_D|--> &= \frac{\mu_0}{4\pi}\gamma^2\hbar^2\overline{\frac{1}{r^3_{12}}\left(1-3\frac{r^2_{12,z}}{r^2_{12}}\right)}\left(\frac{1}{2}\right)^2. \end{aligned} \tag{5.19}$$

For the degenerate states it is necessary to evaluate the eigenvalues and eigenvectors of $\bar{\mathcal{H}}_1$ in the subspace spanned by those vectors $\{|+->, |-+>\}$, this is a subspace of dimension 2 and the matrix elements of $\bar{\mathcal{H}}_1$ in that subspace are

$$
\begin{aligned}
< +-|\bar{\mathcal{H}}_1|+- > &= 2\pi\hbar\left(\bar{J}^{\,s}_{12,zz}+\frac{1}{3}J^{\,tr}_{12}\right)\left(-\frac{1}{4}\right)\\
&\quad+\frac{\mu_0}{4\pi}\gamma^2\hbar^2\overline{\frac{1}{r^3_{12}}\left(1-3\frac{r^2_{12,z}}{r^2_{12}}\right)}\left(-\frac{1}{4}\right)\\
< +-|\bar{\mathcal{H}}_1|-+ > &= 2\pi\hbar\left(-\frac{1}{4}\bar{J}^{\,s}_{12,zz}+\frac{1}{6}J^{\,tr}_{12}\right)\\
&\quad+\frac{\mu_0}{4\pi}\gamma^2\hbar^2\overline{\frac{1}{r^3_{12}}\left(1-3\frac{r^2_{12,z}}{r^2_{12}}\right)}\left(-\frac{1}{4}\right)\\
< -+|\bar{\mathcal{H}}_1|+- > &= < +-|\bar{\mathcal{H}}_1|-+ >\\
< -+|\bar{\mathcal{H}}_1|-+ > &= < +-|\bar{\mathcal{H}}_1|+- > . \qquad (5.20)
\end{aligned}
$$

Solving the eigenvalue equation $\bar{\mathcal{H}}_1\,(C_1|+- > +C_2|-+ >) = E\,(C_1|+- > +C_2|-+ >)$ leads to the eigenvalues

$$
\begin{aligned}
E^1_{0+} &= < +-|\bar{\mathcal{H}}_1|+- > + < +-|\bar{\mathcal{H}}_1|-+ >\\
&= 2\pi\hbar\left(-\frac{1}{2}\bar{J}^{\,s}_{12,zz}+\frac{1}{12}J^{\,tr}_{12}\right)-\frac{1}{2}\frac{\mu_0}{4\pi}\gamma^2\hbar^2\overline{\frac{1}{r^3_{12}}\left(1-3\frac{r^2_{12,z}}{r^2_{12}}\right)}\\
E^1_{0-} &= < +-|\bar{\mathcal{H}}_1|+- > - < +-|\bar{\mathcal{H}}_1|-+ >\\
&= -\frac{\pi}{2}\hbar J^{\,tr}_{12} \qquad (5.21)
\end{aligned}
$$

and the eigenvectors

$$
\begin{aligned}
|\psi_+ > &= 1/\sqrt{2}\,(|+- > +|-+ >)\\
|\psi_- > &= 1/\sqrt{2}\,(|+- > -|-+ >) . \qquad (5.22)
\end{aligned}
$$

The energy of the different states of the total Hamiltonian $\mathcal{H}$ in first-order perturbation is then

$$
\begin{aligned}
E_+ = E^0_+ + E^1_+ &= -\gamma B_0\hbar(1-\bar{\sigma}_{zz})\\
&\quad+2\pi\hbar\left(\bar{J}^{\,s}_{12,zz}+\frac{1}{3}J^{\,tr}_{12}\right)\left(\frac{1}{4}\right)\\
&\quad+\frac{\mu_0}{4\pi}\gamma^2\hbar^2\overline{\frac{1}{r^3_{12}}\left(1-3\frac{r^2_{12,z}}{r^2_{12}}\right)}\left(\frac{1}{4}\right)
\end{aligned}
$$

$$E_{0+} = E_0^0 + E_{0+}^1 = 2\pi\hbar\left(-\frac{1}{2}\bar{J}^{s}_{12,zz} + \frac{1}{12}J^{tr}_{12}\right) - \frac{1}{2}\frac{\mu_0}{4\pi}\gamma^2\hbar^2\overline{\frac{1}{r_{12}^3}\left(1 - 3\frac{r_{12,z}^2}{r_{12}^2}\right)}$$

$$E_{0-} = E_0^0 + E_{0-}^1 = -\frac{\pi}{2}\hbar J^{tr}_{12}$$

$$E_- = E_-^0 + E_-^1 = \gamma B_0\hbar(1 - \bar{\sigma}_{zz}) + 2\pi\hbar\left(\bar{J}^{s}_{12,zz} + 1/3J^{tr}_{12}\right)\left(\frac{1}{2}\right)^2 + \frac{\mu_0}{4\pi}\gamma^2\hbar^2\overline{\frac{1}{r_{12}^3}\left(1 - 3\frac{r_{12,z}^2}{r_{12}^2}\right)}\left(\frac{1}{2}\right)^2 \qquad (5.23)$$

The state $|\psi_- > = 1/\sqrt{2}\,(|+- > -|-+ >)$ is antisymmetric with respect to the exchange of particles while all other states are symmetric. In NMR the $\mathbf{B}_1$ induced transitions between the different energy states can only occur when both the selection rule $\Delta m = \pm 1$ and the conservation of parity are respected, consequently there are no allowed transitions involving state $|\psi_- >$. The allowed transitions take place between the states $|++ >$, $1/\sqrt{2}(|+- > +|-+ >)$ and $|-- >$. Due to the selection rule $\Delta m = \pm 1$ only two distinct transitions can occur, respectively between the states $|++ >$ and $1/\sqrt{2}(|+- > +|-+ >)$ and between the states $1/\sqrt{2}(|+- > +|-+ >)$ and $|-- >$. The NMR signal spectrum is then composed of two lines with angular frequencies ω_+ and ω_- arising from the two allowed transitions. The frequencies take the values

$$\omega_2 = (E_{0+} - E_+)/\hbar = \gamma B_0(1 - \bar{\sigma}_{zz}) - 2\pi\frac{3}{4}\bar{J}^{s}_{12,zz} - \frac{3}{4}\frac{\mu_0}{4\pi}\gamma^2\hbar\overline{\frac{1}{r_{12}^3}\left(1 - 3\frac{r_{12,z}^2}{r_{12}^2}\right)}$$

$$\omega_1 = (E_- - E_{0+})/\hbar = \gamma B_0(1 - \bar{\sigma}_{zz}) + 2\pi\frac{3}{4}\bar{J}^{s}_{12,zz} + \frac{3}{4}\frac{\mu_0}{4\pi}\gamma^2\hbar\overline{\frac{1}{r_{12}^3}\left(1 - 3\frac{r_{12,z}^2}{r_{12}^2}\right)}. \qquad (5.24)$$

It is possible to see from Eq. 5.24 that the two frequencies are equally separated from the central frequency $\gamma B_0(1 - \bar{\sigma}_{zz})$ and the frequency splitting between the two lines given by

$$\Delta\omega = \omega_1 - \omega_2 = \frac{3}{2}\frac{\mu_0}{4\pi}\gamma^2\hbar\overline{\frac{1}{r_{12}^3}\left(1 - 3\frac{r_{12,z}^2}{r_{12}^2}\right)} + 2\pi\frac{3}{2}\bar{J}^s_{12,zz} \quad (5.25)$$

is independent of $\bar{\sigma}_{zz}$. The line splitting in Eq. 5.25 gives a measurement of the sum of the zz components of the dipolar and indirect spin–spin coupling averaged interaction tensors, if they are known at the molecular level this result can be used to determine order parameters, as discussed in 5.3. $\bar{\sigma}_{zz}$ could in principle be used for the same purpose but the experimental chemical shift is also affected by the magnetic susceptibility of the sample and the NMR probe head, making this task difficult. The amplitude of the two spectral lines corresponding to the allowed transitions, respectively, W_1 and W_2, can be obtained from Eq. 4.31 which gives the transition probability per unit time induced by the perturbing Hamiltonian arising from the RF field, leading to

$$W_1 \propto | < - - |(I_{1,-} + I_{2,-})\frac{1}{\sqrt{2}}(| - + > + | + - >)|^2 = 2$$

$$W_2 \propto |\frac{1}{\sqrt{2}}(< - + | + < + - |)(I_{1,-} + I_{2,-})| + + > |^2 = 2 \quad (5.26)$$

and showing that the two lines have the same amplitude, the spectrum is symmetric around the frequency $\gamma B_0(1 - \bar{\sigma}_{zz})$.

5.4.2 NMR Spectrum from One Spin 1 Particle Subjected to Quadrupolar Interaction and the Chemical Shift

The spin 1 particle system subjected to the quadrupolar interaction is particularly interesting when dealing with a deuterium atom in a CD bond, this is a system that is used in NMR of anisotropic fluids through selective deuteration to look at specific sites in a molecule. The unperturbed Hamiltonian is composed of the Zeeman Hamiltonian

$$\mathcal{H}_0 = \mathcal{H}_z = -\gamma\hbar B_0 I_z \quad (5.27)$$

and the perturbation

$$\bar{\mathcal{H}}_1 = \bar{\mathcal{H}}_\sigma + \bar{\mathcal{H}}_Q = \gamma\hbar I_z\bar{\sigma}_{zz}B_0 + \frac{3eQ}{4}\left[I_z^2 - \frac{2}{3}\right]\bar{V}_{zz}. \quad (5.28)$$

In accord with the method outlined in Chapter 4, Section 4.4.4.1, the eigenstates of the unperturbed Hamiltonian $\mathcal{H}_0 = \mathcal{H}_z$ are eigenstates $|I, m>$ of the operators $(\mathbf{I})^2$ and I_z. The Hilbert space for this system has dimension 3 and the three base states which are eigenstates of the unperturbed Hamiltonian are explicitly

$$|1, 1>, |1, 0>, |1, -1> \tag{5.29}$$

and will be abbreviated to $|+>, |0>, |->$, respectively. The unperturbed energy associated with each state is

$$\begin{aligned} E_+^0 &= <+|\mathcal{H}_z|+> = -\gamma\hbar B_0 <+|I_z|+> = -\gamma B_0\hbar \\ E_0^0 &= <0|\mathcal{H}_z|0> = -\gamma\hbar B_0 <0|I_z|0> = 0 \\ E_-^0 &= <-|\mathcal{H}_z|-> = -\gamma\hbar B_0 <-|I_z|-> = \gamma B_0\hbar. \end{aligned} \tag{5.30}$$

The first-order energy corrections arising from the perturbing Hamiltonian $\bar{\mathcal{H}}_1$ are

$$\begin{aligned} E_+^1 &= <+|\bar{\mathcal{H}}_1|+> = \gamma B_0\hbar\bar{\sigma}_{zz} + \frac{eQ}{4}\bar{V}_{zz} \\ E_0^1 &= <0|\bar{\mathcal{H}}_1|0> = -\frac{eQ}{2}\bar{V}_{zz} \\ E_-^1 &= <-|\bar{\mathcal{H}}_1|-> = -\gamma B_0\hbar\bar{\sigma}_{zz} + \frac{eQ}{4}\bar{V}_{zz} \end{aligned} \tag{5.31}$$

leading to the energy levels of Hamiltonian

$$\begin{aligned} E_+ &= -\gamma B_0\hbar(1-\bar{\sigma}_{zz}) + \frac{eQ}{4}\bar{V}_{zz} \\ E_0 &= -\frac{eQ}{2}\bar{V}_{zz} \\ E_- &= \gamma B_0\hbar(1-\bar{\sigma}_{zz}) + \frac{eQ}{4}\bar{V}_{zz}. \end{aligned} \tag{5.32}$$

The allowed transitions occur between states $|1, 1>$ and $|1, 0>$ and between states $|1, 0>$ and $|1, -1>$ given rise to an NMR signal with a two lines spectrum with angular frequencies

$$\begin{aligned} \omega_1 &= \Delta E_1/\hbar = (E_- - E_0)/\hbar = \gamma B_0(1-\bar{\sigma}_{zz}) + \frac{3eQ}{4\hbar}\bar{V}_{zz}. \\ \omega_2 &= \Delta E_2/\hbar = (E_0 - E_+)/\hbar = \gamma B_0(1-\bar{\sigma}_{zz}) - \frac{3eQ}{4\hbar}\bar{V}_{zz} \end{aligned} \tag{5.33}$$

The frequency splitting between the two lines is a direct measure of the averaged *zz* component of the electric field gradient tensor at the nucleus site

$$\Delta\omega = \omega_1 - \omega_2 = \frac{3eQ}{2\hbar}\bar{V}_{zz}. \tag{5.34}$$

The amplitude of the two spectral lines, respectively W_1 and W_2, can be obtained from Eq. 4.31 leading to

$$W_1 \propto | < -|I_-|0 > |^2 = 2$$
$$W_2 \propto | < 0|I_-|+ > |^2 = 2 \tag{5.35}$$

which shows that the spectrum is symmetric around the frequency $\gamma B_0(1 - \bar{\sigma}_{zz})$.

5.5 Quantum Mechanical Analysis of Selected NMR Pulse Sequences

NMR pulse sequences play a central role in the modern NMR technique; they allow the manipulation of the nuclear spin Hamiltonian with great versatility, setting up the conditions necessary for the desired NMR observable to be detected. The pulse sequences to be discussed next include the one pulse sequence, which, as discussed in Section 4.6 for a $\pi/2$ pulse, gives rise to a *free induction decay* (FID) whose Fourier transform coincides with the spin Hamiltonian absorption spectrum, and the solid echo pulse sequence which is quite frequently used in 1D high-field NMR spectroscopy of anisotropic fluids.

5.5.1 Spin System Subjected to a Single $\pi/2$ RF Pulse in Resonance

In high-field pulsed NMR spectroscopy the quantity experimentally recorded is the complex transverse magnetization $m(t) = M_x(t) + i M_y(t)$ (4.50), whose components in the lab frame $[x, y, z]$ are related to those in the rotating frame $[x', y', z']$ (4.3.3) by

$$M_x(t) + i M_y(t) = \left(M_{x'}(t) + i M_{y'}(t)\right) e^{-i\omega t} = N\gamma\hbar Tr\left\{\bar{\rho}'(t) I^+\right\} e^{-i\omega t} \tag{5.36}$$

where $\bar{\rho}'(t)$ is the density matrix in the rotating frame, $\omega = \gamma B_0$, and N the number density of target spins. Knowing $\bar{\rho}'(t)$ enables the determination of $m(t)$ through Eq. 5.36. As discussed previously in 4.6.2 it is more convenient to analyze the evolution of the spin system in the rotating frame where the time evolution of the density matrix is simplified. In order to carry out explicitly all the calculations, an ensemble of spin 1 particles subjected to an axially symmetric quadrupolar interaction is considered in the analysis. As the spins are not interacting with each other, it is sufficient to consider just one spin. The time evolution of the density matrix starts at $t = 0^-$ with the spin system in thermal equilibrium with the lattice, leading to the following density matrix in the rotating frame

$$\begin{aligned}\bar{\rho}'(0^-) &= R(0)\bar{\rho}(0^-)R^{-1}(0) = \bar{\rho}(0^-) = \frac{1}{Z}e^{-\mathcal{H}/kT} \\ &\cong \frac{1}{Z}(1 - \mathcal{H}/kT) \\ &= \frac{1}{Z}\left\{1 - \frac{1}{kT}\left[-\gamma\hbar B_0 I_z + \frac{3eQ}{4}\left(I_z^2 - \frac{2}{3}\right)\bar{V}_{zz}\right]\right\}.\end{aligned} \quad (5.37)$$

At time $t = 0$ the RF B_1 field is switched on for a short period of time sufficient to tilt the magnetization by 90°. In the rotating frame the time evolution of the density matrix is governed by Eq. 4.53 and while the B_1 field is on the effective Hamiltonian is $\mathcal{H}_e \approx -\gamma\hbar B_1 I_x$ and the density matrix follows a solution of the type given in 4.44 leading to

$$\bar{\rho}'(t) = e^{\{i\omega_1 I_x t_1\}}\bar{\rho}'(0^-)e^{\{-i\omega_1 I_x t\}}, \quad (5.38)$$

where $\omega_1 = \gamma B_1$. Using the following relation obeyed by the spin operators (Kimmich, 1997)

$$e^{\{-i\phi I_x\}}I_z e^{\{i\phi I_x\}} = I_z\cos(\phi) - I_y\sin(\phi) \quad (5.39)$$

and considering in Eq. 5.37 that the quadrupolar term in the Hamiltonian is much smaller than the Zeeman term, Eq. 5.38 gives the result

$$\bar{\rho}'(t) = \frac{1}{Z}\left\{1 + \frac{\gamma\hbar B_0}{kT}\left[I_z\cos(\omega_1 t) + I_y\sin(\omega_1 t)\right]\right\}. \quad (5.40)$$

For $t = t_1 \equiv \frac{\pi}{2\omega_1}$ we get

$$\bar{\rho}'(t = t_1) = \frac{1}{Z}\left[1 + \frac{\gamma\hbar B_0}{kT}I_y\right]. \quad (5.41)$$

At $t = t_1 = \frac{\pi/2}{\omega_1}$ the RF pulse is switched off and the density matrix enters a new evolution time period with the effective Hamiltonian taking the form

$$\mathcal{H}_e = R\mathcal{H}_i R^{-1} = e^{\{-i\omega t I_z\}}\left[\frac{3eQ}{4}\left(I_z^2 - \frac{2}{3}\right)\bar{V}_{zz}\right]e^{\{i\omega t I_z\}}$$
$$= \frac{3eQ}{4}\left(I_z^2 - \frac{2}{3}\right)\bar{V}_{zz}. \quad (5.42)$$

Again due to the simplifying assumptions considered the effective Hamiltonian is time independent and a solution of the type 4.44 is verified leading to

$$\bar{\rho}'(t_2) = e^{\{\frac{-i}{\hbar}\mathcal{H}_e t_2\}}\bar{\rho}'\left(t_1 = \frac{\pi/2}{\omega_1}\right)e^{\{\frac{i}{\hbar}\mathcal{H}_e t_2\}}. \quad (5.43)$$

where $t_2 = t - t_1$. Using the results of Eq. 5.41 and Eq. 5.42, $\bar{\rho}'$ becomes

$$\bar{\rho}'(t_2) = \frac{1}{Z}\left[1 + \frac{\gamma\hbar B_0}{kT}e^{\{\frac{-i}{\hbar}\frac{3eQ}{4}(I_z^2-\frac{2}{3})\bar{V}_{zz}t_2\}}I_y e^{\{\frac{i}{\hbar}\frac{3eQ}{4}(I_z^2-\frac{2}{3})\bar{V}_{zz}t_2\}}\right] \quad (5.44)$$

and using the relation obeyed by the spin 1 operators (Kimmich, 1997)

$$e^{\{-i\phi[I_z^2-\frac{2}{3}]\}}I_y e^{\{i\phi[I_z^2-\frac{2}{3}]\}} = I_y\cos(\phi) - (I_x I_z + I_z I_x)\sin(\phi) \quad (5.45)$$

one obtains

$$\bar{\rho}'(t_2) = \frac{1}{Z}\left\{1 + \frac{\gamma\hbar B_0}{kT}\left[I_y\cos\left(\frac{3eQ}{4\hbar}\bar{V}_{zz}t_2\right)\right.\right.$$
$$\left.\left. - (I_x I_z + I_z I_x)\sin\left(\frac{3eQ}{4\hbar}\bar{V}_{zz}t_2\right)\right]\right\}. \quad (5.46)$$

Inserting Eq. 5.46 in Eq. 5.36 one obtains for the complex transverse magnetization

$$M_x(t) + iM_y(t) = N\gamma\hbar Tr\left\{\bar{\rho}'(t)I^+\right\}e^{-i\omega t}$$
$$= \frac{N\gamma^2\hbar^2 B_0}{KT}2i\cos\left[\frac{3eQ}{4\hbar}\bar{V}_{zz}(t - t_1)\right]e^{-i\omega t} \quad (5.47)$$

The complex magnetization given by Eq. 5.47 has a Fourier transform composed of two delta functions at the frequencies $\gamma B_0 + \frac{3eQ}{4\hbar}\bar{V}_{zz}$ and $\gamma B_0 - \frac{3eQ}{4\hbar}\bar{V}_{zz}$ which are precisely the absorption frequencies of the spin Hamiltonian considered (see Eq. 5.33), in agreement with the discussion of the method in 4.6. The fact that the result Eq. 5.47 gives a transverse magnetization that oscillates but does not decay is a consequence of the simplified Hamiltonian considered in Eq. 5.37 that excludes relaxation.

5.5.2 Solid Echo Pulse Sequence

The solid echo pulse sequence is composed of two $\pi/2$ RF pulses, the first one along the X' axis of the rotating frame and the second one along the Y' axis. This pulse sequence generates an echo on the detected signal that appears after the second pulse and separated from it of a time interval equal to the pulse separation. The signal starting from the echo top onward reproduces the FID that is generated by a single $\pi/2$ pulse. The use of this pulse sequence in the determination of the absorption spectra of the spin Hamiltonian presents advantages over the single $\pi/2$ pulse sequence specially for wide-frequency-range spectra associated to fast FIDs. After a $\pi/2$ RF pulse the signal detecting electronics stays nonoperational for some microseconds and if the FID is fast this loss of signal is not acceptable. With the solid echo pulse sequence the signal to be Fourier transformed starts at the echo top that may be placed sufficiently apart from the RF pulses allowing the recording of the FID without any data loss. To carry out all the calculations the same spin system considered for the one pulse sequence will be used. The effect of the first $\pi/2$ RF pulse along the X' axis on the density matrix was determined before giving rise to density matrix reported in Eq. 5.46 that at $t_2 = \tau$ where τ is the pulse separation takes the value

$$\bar{\rho}'(t_2 = \tau) = \frac{1}{Z}\left\{1 + \frac{\gamma\hbar B_0}{kT}\left[I_y \cos\left(\frac{3eQ}{4\hbar}\bar{V}_{zz}\tau\right) - (I_x I_z + I_z I_x)\sin\left(\frac{3eQ}{4\hbar}\bar{V}_{zz}\tau\right)\right]\right\}. \quad (5.48)$$

At $t_2 = \tau$ a $\pi/2$ RF pulse along Y' is applied to the spin system, this is achieved with a radio-frequency (RF) term in the Hamiltonian of the same form as used for the X' pulse but dephased by $\pi/2$ as follows

$$\mathcal{H}_{RF} = -2\gamma\hbar I_x B_1 \cos(\omega t - \pi/2) \quad (5.49)$$

and the contribution to effective Hamiltonian in the rotating frame at resonance becomes

$$\begin{aligned} R\mathcal{H}_{RF}R^{-1} &= R[-2\gamma\hbar I_x B_1 \cos(\omega t - \pi/2)]R^{-1} \\ &= -\gamma\hbar B_1 I_y - \gamma\hbar B_1\left[\sin(2\omega t)I_x - \cos(2\omega t)I_y\right]. \end{aligned} \quad (5.50)$$

Neglecting the oscillating terms that are not effective in changing the magnetization and also the contribution from the interaction terms $\mathcal{H}_i'$ as done for the X' pulse, the effective Hamiltonian while the Y' pulse is ON becomes $\mathcal{H}_e \approx -\gamma\hbar B_1 I_y$. While the Y' pulse is ON the density matrix evolves according to

$$\bar{\rho}'(t_3) = e^{\{i\omega_1 I_y t_3\}}\bar{\rho}'(\tau)e^{\{-i\omega_1 I_y t_3\}}, \tag{5.51}$$

where $\omega_1 = \gamma B_1$ and $t_3 = t - t_1 - t_2$. Using the following relation obeyed by the spin 1 operators (Kimmich, 1997)

$$e^{\{-i\phi I_y\}}\,(I_x I_z + I_z I_x)\,e^{\{i\phi I_y\}} = (I_x I_z + I_z I_x)\cos(2\phi) - (I_z^2 - I_x^2)\sin(2\phi) \tag{5.52}$$

and evaluating $\bar{\rho}'\left(t_3 = \frac{\pi}{2\omega_1}\right)$ which corresponds to a $\pi/2$ Y' pulse leads to the result

$$\bar{\rho}'\left(t_3 = \frac{\pi}{2\omega_1}\right) = \frac{1}{Z}\left\{1 + \frac{\gamma\hbar B_0}{kT}\left[\cos\left(\frac{3eQ}{4\hbar}\bar{V}_{zz}\tau\right)I_y + \sin\left(\frac{3eQ}{4\hbar}\bar{V}_{zz}\tau\right)(I_x I_z + I_z I_x)\right]\right\}. \tag{5.53}$$

At $t_3 = \frac{\pi}{2\omega_1}$ the RF pulse is switched off and the density matrix enters a new evolution period with the effective Hamiltonian given by Eq. 5.42. The solution of the von Neumann equation in the rotating frame in this new period yields the result

$$\bar{\rho}'(t_4) = e^{\{\frac{-i}{\hbar}\mathcal{H}_e t_4\}}\bar{\rho}'\left(t_3 = \frac{\pi}{2\omega_1}\right)e^{\{\frac{i}{\hbar}\mathcal{H}_e t_4\}} \tag{5.54}$$

where $t_4 = t - t_1 - t_2 - t_3$. Using Eq. 5.45 and the following relation obeyed by the spin 1 operators (Kimmich, 1997)

$$e^{\{-i\phi[I_z^2 - \frac{2}{3}]\}}(I_x I_z + I_z I_x)e^{\{i\phi[I_z^2 - \frac{2}{3}]\}} = (I_x I_z + I_z I_x)\cos(\phi) + I_y\sin(\phi), \tag{5.55}$$

the density matrix given by Eq. 5.54 takes the form

$$\bar{\rho}'(t_4) = \frac{1}{Z}\left\{1 + \frac{\gamma\hbar B_0}{kT}\left[I_y\cos\left(\frac{3eQ}{4\hbar}\bar{V}_{zz}(\tau - t_4)\right) + (I_z I_x + I_x I_z)\sin\left(\frac{3eQ}{4\hbar}\bar{V}_{zz}(\tau - t_4)\right)\right]\right\}. \tag{5.56}$$

The transverse magnetization in the time interval after the second pulse becomes

$$\begin{aligned} M_x(t) + iM_y(t) &= N\gamma\hbar Tr\left\{\bar{\rho}'(t)I^+\right\}e^{-i\omega t} \\ &= \frac{N\gamma^2\hbar^2 B_0}{kT}2i\cos\left[\frac{3eQ}{4\hbar}\bar{V}_{zz}(\tau - t_4)\right]e^{-i\omega t}. \end{aligned} \tag{5.57}$$

When $t_4 = \tau$ in Eq. 5.57 the transverse magnetization reaches a maximum value independent of the strength of the quadrupolar interaction given by $\frac{eQ\bar{V}_{zz}}{\hbar}$ showing that an echo forms at this time. The time dependence of the complex magnetization starting from the echo maximum onward is identical to the one obtained after just one $\pi/2$ X' pulse justifying its use in obtaining the Hamiltonian absorption spectra through Fourier transform of the FID. Once more the expected decay of the transverse magnetization is not observed in Eq. 5.57 as the simplified Hamiltonian considered for the spin system excludes relaxation.

5.6 Experimental Details

NMR equipment operates in the RF range that includes the bandwidths of radio stations operating in most countries.[a]

All NMR spectrometers include a magnetic field source (e.g., permanent magnet, conventional or superconductor electromagnet) and an emitter/receiver RF system, usually referred to as the NMR console. Modern NMR setups include computer systems for setup control and data acquisition. One important element of the emitter/receiver unit is the RF antenna that allows for the sample's irradiation and signal reception. In Fig. 5.1 is presented a cartoon that illustrates some main basic features of the coupling between the external magnetic field, the nuclear spin magnetization and the RF in resonance with the Larmor frequency.

The solenoid coil is the emitter/receiver RF antenna inside which the sample is placed.[b] The shape and number of coil windings that define the coil's self-inductance and the high-power capacitors are used to form a RLC resonance electric circuit tuned at the resonance frequency of the emitter/receiver frequency and also the nuclear spins' Larmor frequency. The coil axis is set perpendicular to the main, static Zeeman magnetic field in order for the RF field components (e.g., $\vec{B}_L$) to lie in the plane perpendicular to the Zeeman field. In fact, the RF field is linearly polarized along the

---

[a] For this reason RF shielding is of crucial importance in all NMR equipment.

[b] Depending of the probe head design the RF coils can be solenoidal or saddle coils.

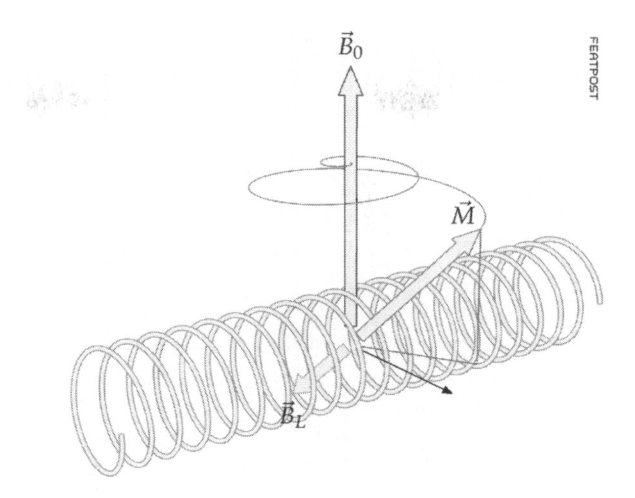

Figure 5.1 Illustrative cartoon of the emitter/receiver coil, the magnetic field, and the coupling between the static magnetic field, the RF field, and the nuclear spin magnetization. The path of the magnetization vector tip illustrates the effect of the applied RF field on the nuclear magnetization for a nucleus with a positive gyromagnetic ratio ($\gamma > 0$). Cartoon generated with FEATPOST software (Gonçalves, 2004).

coil axis. It can be considered as the combination of the rotating fields in opposite directions. The only component in resonance with the Larmor frequency is that rotating in the same direction as the Larmor precession, as described in Chapter 4. According to Bloch's equations (Eq. 4.19) the magnetization precesses around the vertical axis and always perpendicular to the RF field and the angle between the nuclear spin magnetization and the external magnetic field increases with time. If the RF is a pulse with length $t_p = \pi/(2\gamma B_L)$ at the end of the pulse the magnetization will be in the plane perpendicular to $\vec{B}_0$. Such pulse is referred to as a $\pi/2$-*pulse*. An RF pulse with length $2t_p$ will leave the nuclear spin magnetization aligned antiparallel to $\vec{B}_0$ and is referred to as a π-*pulse* (Farrar and Becker, 1971).

The same coil is used to detect the sample's response signal as the result of the nuclear spin relaxation processes as the result of the evolution of the magnetization back to its equilibrium state (e.g., aligned parallel to $\vec{B}_0$), in the absence of any applied RF.

The precession of $\vec{M}$ components perpendicular to $\vec{B}_0$ will induce a electromotive force in the coil and a time-decaying AC current will be generated in the resonance circuit with the Larmor frequency. After signal amplification and rectification the low-frequency components can be analyzed and viewed on an oscilloscope or equivalent. The detected signal is the FID. The time evolution of magnetization components aligned with $\vec{B}_0$ will not be detected in the coil.

The amplitude of the signal detected in the coil depends on several factors (Abragam, 1961). Important factors are the quality factor of the resonance electrical circuit, the value of B_0, and the filling factor (e.g., the *volume of the sample/volume of the coil* ratio). For a good experiment it is important to adjust the system to obtain the largest signal amplitude possible. The *signal/noise* ratio can be a limiting condition in any NMR experiment. In particular, this ratio depends on $B_0^{3/2}$ and for magnetic fields below 0.2 T is it hard to obtain good NMR FIDs (Abragam, 1961; Noack, 1986).

Due to its gyromagnetic ratio hydrogen proton spin has the largest Larmor frequency for a given magnetic field in comparison with all other nuclear spins. For a 7.05 T magnetic field the ^{1}H Larmor frequency corresponds to approximately 300 MHz.[a]

[a]The ^{1}H Larmor frequency for 1 T is 42.577 MHz.

Chapter 6

NMR Relaxation and Molecular Dynamics: Theory

6.1 General Concepts

The fact that a system of nuclear spins interacts with magnetic fields and is able to absorb and dissipate energy is a fundamental aspect of nuclear magnetic resonance (NMR). The energy exchange within the nuclear spin system and between the system and the surrounding environment is commonly referred to as relaxation and is one very important problem in NMR.

The concept of relaxation is closely associated with the concept of evolution of populations in the different energy levels. When in the steady state the population of the spin system energy levels follows a Boltzmann distribution. In the case of spins $I = 1/2$ the ratio between the populations of the high and low energy levels, p_+ and p_-, respectively, with $p_+ + p_- = 1$, is given by $p_+/p_- = \exp(\hbar\gamma B_0/(kT))$, where T is the temperature of the sample. The interaction of the spin systems with electromagnetic waves with frequency $\omega = \omega_0 = \hbar\gamma B_0$ (i.e., equal to the Larmor frequency) disturbs the equilibrium state, and a new equilibrium results from the competition between the radio-frequency (RF) irradiation which

NMR of Liquid Crystal Dendrimers
Carlos R. Cruz, João L. Figueirinhas, and Pedro J. Sebastião

ISBN 978-981-4745-72-7 (Hardcover), 978-981-4745-73-4 (eBook)
www.panstanford.com

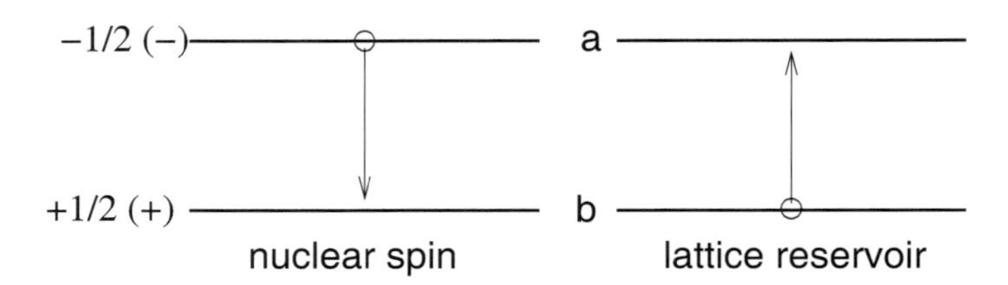

Figure 6.1 Energy levels of an $I = 1/2$ spin system and a lattice reservoir, both with the same energy gap.

promotes the increase of p_+/p_- and the relaxation process which tends to decrease this ratio toward the initial value. A new state is reached with $p_+/p_- = \exp(\hbar\gamma B_0/(kT_S))$ when the RF irradiation stops and the system relaxes toward the original equilibrium. T_S in the Boltzmann factor represents the new "temperature" of the spin system at the end of the RF pulse and it is clear that $T_S > T$. This definition of "spin temperature" can be extended also to spin systems with $I > 1/2$. This concept is not strictly valid when the RF is being applied, since there are transverse components of the magnetization which are not compatible with the description of the statistical behavior of spin systems by the population of its energy levels. One important aspect of the "spin temperature" concept is related with the saturation state when $p_+ = p_-$, which corresponds to an "infinite temperature" case. Surprising is the case observed when $p_+ < p_-$, which leads to the concept of a "negative." This situation is observed when a 180° pulse is applied and the nuclear magnetization is inverted. Obviously, a "negative" T_S corresponds to a "hotter" state in comparison with the "infinite temperature" state (Abragam, 1961).

The balance between the populations in the two energy levels of $I = 1/2$ spin systems can be used also to introduce the concept of relaxation time. In the relaxation process the rate of change of one energy level is obviously coupled with the rate of change of the other energy level. For the $I = +1/2$ we have an energy $E_+ = -(1/2)\hbar\gamma B_0 p_+$ and for $I = -1/2$ we have $E_- = +(1/2)\hbar\gamma B_0 p_-$. For an isolated spin system the rate of change of p_+ can be written as

$$\frac{dp_+}{dt} = W_\downarrow p_- - W_\uparrow p_+ \tag{6.1}$$

and clearly the transition probabilities per unit of second for the transitions $(-) \rightarrow (+)$ and $(+) \rightarrow (-)$, respectively, $W_{\downarrow}$ and $W_{\uparrow}$, must be equal, as it can be demonstrated by the time-dependent perturbation theory in quantum mechanics (Slichter, 1992).

In the case a nuclear spin system coupled with the latticed reservoir only the transitions $(-) \rightarrow (+)$ with $(a) \rightarrow (b)$, with probability per second $W_{\downarrow\uparrow}$ and $(+) \rightarrow (-)$ with $(b) \rightarrow (a)$, with probability per second $W_{\uparrow\downarrow}$ are allowed according to basic quantum mechanical principles. The total number of transitions per second in a steady state gives $p_- p_b W_{\downarrow\uparrow} = p_+ p_a W_{\uparrow\downarrow}$ and since $W_{\downarrow\uparrow} = W_{\uparrow\downarrow}$ it happens that $p_-/p_+ = p_a/p_b$, where p_a and p_b are the populations of the (a) and (b) energy levels of the lattice reservoir.

At equilibrium the total number of transitions per second can also be written in the form $p_- W_{\downarrow} = p_+ W_{\uparrow}$ with $W_{\downarrow} = p_b W_{\downarrow\uparrow}$ and $W_{\uparrow} = p_a W_{\uparrow\downarrow}$. Since $W_{\downarrow\uparrow} = W_{\uparrow\downarrow}$ the upward and downward transition probabilities per second for the spin systems coupled with the lattice are different.

The rate of change of the difference between the population in the two spin states $n = p_+ - p_-$ can be written as

$$\frac{dn}{dt} = \frac{n_0 - n}{T_1} \tag{6.2}$$

with $n_0 = (W_{\downarrow} - W_{\uparrow})/(W_{\downarrow} + W_{\uparrow})$ and $T_1 = W_{\downarrow} + W_{\uparrow}$ (Slichter, 1992). T_1 is the *spin–lattice relaxation* time and its inverse, T_1^{-1}, is *spin–lattice relaxation rate.*

The quantification of the spin–lattice relaxation rate, as well as the calculation of other relaxation rates that will be introduced below are only possible within the quantum mechanics formalism, in particular within the time-dependent perturbation theory. One often used theoretical approach is that of the Bloch–Wangsness–Redfield (BWR) density operator perturbation for weak collisions (Abragam, 1961). The spin Hamiltonian is written as

$$\mathcal{H}(t) = \mathcal{H}_0 + \mathcal{H}_1(t) \tag{6.3}$$

where $\mathcal{H}_0$ is the static part of the Hamiltonian and $\mathcal{H}_1(t)$ is the perturbation Hamiltonian usually associated with the spin interactions.

According to this theory the density operator of the spin system σ can be used to estimate the expectation value of any observable Q

of the system according with the usual quantum theory expression

$$\langle Q \rangle = Tr\{\sigma(t)Q\}. \tag{6.4}$$

For the calculus of the density operator is necessary to solve the differential equation

$$\frac{d\sigma(t)}{dt} = -\frac{i}{\hbar}[\mathcal{H}_0 + \mathcal{H}_1(t), \sigma] \tag{6.5}$$

This very formal expression is more often used in a more practical form in the interaction representation of the operators. Using the unitary quantum operator $\mathcal{U} \equiv \exp(-(i/h)\mathcal{H}_0 t)$, and the relations $\mathcal{H}_0 = \mathcal{U}\mathcal{H}'_0\mathcal{U}^{-1} = \mathcal{H}'_0$, $\mathcal{H}_1 = \mathcal{U}\mathcal{H}'_1\mathcal{U}^{-1}$, and $\sigma' = \mathcal{U}\sigma\mathcal{U}^{-1}$ it is possible to write

$$\overline{\frac{d\sigma'(t)}{dt}} = -\frac{1}{\hbar^2}\int_0^\infty \overline{\left[\mathcal{H}'_1(t), \left[\mathcal{H}'_1(t-\tau), \sigma'(0)\right]\right]}d\tau \tag{6.6}$$

valid for $t \gg \tau_c$ where τ_c is the characteristic time of the spin interactions (Slichter, 1992) and for a stochastically fluctuation perturbation $\overline{\mathcal{H}'(t)} = 0$.

The perturbation Hamiltonian corresponding to the spin interactions most often considered in NMR relaxation, can be written the very compact form

$$H_\lambda(t) = K_\lambda \sum_{m=-2}^{2} T_{2,m} F_{2,m}(\vec{r}(t)) \tag{6.7}$$

where K_λ is the interaction strength. $T_{2,m}$ represents the spin part of the Hamiltonian terms, and $F_{2,m}(\vec{r}(t))$ the spatial part of the time fluctuating Hamiltonian.

Since the hydrogen nucleus has spin 1/2 and hydrogen is the most abundant chemical element in organic materials, the following theoretical treatment will focus on the dipolar interaction between spins. For ^{1}H spins the remaining spin interactions are less relevant for NMR relaxation studies.

6.2 Relaxation Rates

The spin–lattice (or *longitudinal*) and spin–spin (or *transverse*) relaxation rates T_1^{-1} and T_2^{-1}, respectively, are calculated taking into account the most important spin interaction. The direct dipolar coupling and the quadrupolar coupling are undoubtedly two of the most common and relevant in NMR relaxation. Considering the most common case where the direct dipolar coupling is between two spins (e.g., two ^{1}H spins with $I = 1/2$) here labeled I and S, respectively, the Hamiltonian can be written as (Abragam, 1961; Kimmich, 1997)

$$H_{dd}(t) = K_{dd} \sum_{m=-2}^{2} A^{(m)} F^{(m)}(\vec{r}(t)) \tag{6.8}$$

with

$$K_{dd} = \frac{\mu_0}{4\pi} \gamma_I \gamma_S \hbar, \tag{6.9}$$

$$\begin{aligned} A^{(-2)} &= -\frac{3}{4} I^- S^- \\ A^{(-1)} &= -\frac{3}{2}(I^- S_z - +I_z S^-) \\ A^{(0)} &= I_z S_z - \frac{1}{4}(I^+ S^- + I^- S^+) \\ A^{(1)} &= -\frac{3}{2}(I^+ S_z - +I_z S^+) \\ A^{(2)} &= -\frac{3}{4} I^+ S^+ \end{aligned} \tag{6.10}$$

and

$$\begin{aligned} F^{(0)} &= (1 - 3\cos^2\theta) r^{-3} = -\sqrt{\frac{16\pi}{5}} \frac{1}{r^3} Y_2^{(0)}(\theta, \varphi) \\ F^{(1)} &= F^{(-1)*} = \sin\theta \cos\theta e^{-i\varphi} r^{-3} = \sqrt{\frac{8\pi}{15}} \frac{1}{r^3} Y_2^{(-1)}(\theta, \varphi) \\ F^{(2)} &= F^{(-2)*} = \sin^2\theta e^{-2i\varphi} r^{-3} = \sqrt{\frac{32\pi}{15}} \frac{1}{r^3} Y_2^{(-2)}(\theta, \varphi) \end{aligned} \tag{6.11}$$

θ and φ are the Euler angles between the interspin vector in the laboratory frame, and $Y_l^{(m)}(\theta, \varphi)$ are spherical harmonics (Kimmich, 1997).

Considering the static Zeeman Hamiltonian for the two spins

$$\mathcal{H}_z = -\hbar\gamma_I I_z - \hbar\gamma_S S_z \tag{6.12}$$

the expected values of $\langle I_z \rangle$ and $\langle S_z \rangle$ can be calculates using Eq. 6.4 and the solution of Eq. 6.5. This calculus can be found in different text books (Abragam, 1961; Kimmich, 1997; Slichter, 1992) and the results is

$$\frac{d\langle I_z\rangle}{dt} = -\frac{\langle I_z\rangle - \langle I_z\rangle_0}{T_1^{II}} - \frac{\langle S_z\rangle - \langle S_z\rangle_0}{T_1^{IS}} \tag{6.13}$$

$$\frac{d\langle S_z\rangle}{dt} = -\frac{\langle I_z\rangle - \langle I_z\rangle_0}{T_1^{SI}} - \frac{\langle S_z\rangle - \langle S_z\rangle_0}{T_1^{SS}} \tag{6.14}$$

where $\langle I_z\rangle_0$ and $\langle S_z\rangle_0$ are the equilibrium values, and

$$(T_1^{II})^{-1} = K_{dd}^2 S(S+1)\left[\frac{1}{12}J^{(0)}(\omega_I - \omega_S) + \frac{3}{2}J^{(1)}(\omega_I) + \frac{3}{4}J^{(2)}(\omega_I + \omega_S)\right] \tag{6.15}$$

$$(T_1^{IS})^{-1} = K_{dd}^2 I(I+1)\left[-\frac{1}{12}J^{(0)}(\omega_I - \omega_S) + \frac{3}{4}J^{(2)}(\omega_I + \omega_S)\right] \tag{6.16}$$

$$(T_1^{SI})^{-1} = K_{dd}^2 S(S+1)\left[-\frac{1}{12}J^{(0)}(\omega_I - \omega_S) + \frac{3}{4}J^{(2)}(\omega_I + \omega_S)\right] \tag{6.17}$$

$$(T_1^{SS})^{-1} = K_{dd}^2 I(I+1)\left[\frac{1}{12}J^{(0)}(\omega_I - \omega_S) + \frac{3}{2}J^{(1)}(\omega_I) + \frac{3}{4}J^{(2)}(\omega_I + \omega_S)\right] \tag{6.18}$$

Using the Bloch/Wangsness/Redfield theory (Redfield, 1965) it is possible to obtain the expressions for the spin–lattice and spin–spin relaxation rates in terms of the spectral densities, J_α (Abragam, 1961; Kimmich, 1997; Kowalewski and Mäler, 2006).

$$J^{(\alpha)}(\omega_0) = \int_0^\infty G^{(\alpha)}(t)\exp(-i\omega_0 t)dt\ . \tag{6.19}$$

calculated from the autocorrelation functions, $G(t)$,

$$G^{(\alpha)}(t) = \overline{F^{(\alpha)}(t')F^{(\alpha)*}(t+t')} \tag{6.20}$$

For two identical spins $I = S$, we can expect $\langle I_z\rangle = \langle S_z\rangle$, $\langle I_z\rangle_0 = \langle S_z\rangle_0$, and $\omega_I = \omega_S = \omega_0$. Therefore, Eq. 6.13 and Eq. 6.14 merge and give

$$\frac{d\langle I_z + S_z\rangle}{dt} = -\frac{\langle I_z + S_z\rangle - \langle I_z + S_z\rangle_0}{T_1} \tag{6.21}$$

with $1/T_1 = 1/T_1^{II} + 1/T_1^{IS}$ and

$$\frac{1}{T_1} = \frac{3}{2}K_{dd}^2 I(I+1)\left[J^{(1)}(\omega_0) + J^{(2)}(2\omega_0)\right] \tag{6.22}$$

Obviously, the number of identical spins in a sample per cubic centimeter is in the order of 10^{23} (i.e., of the order of the Avogadro's number), therefore for the calculus of the spin–lattice relaxation rate there are two cases to consider: (*i*) the motion of the identical spins are correlated or (*ii*) the motions are uncorrelated.

(1) If the spins are correlated then to a very good approximation a single exponential decay evolution is observed for the spin–lattice relaxation with a single relaxation rate where the spectral densities Eq. 6.19 are calculated using ensemble averages of the autocorrelation functions

$$\left\langle G^{(\alpha)}(t)\right\rangle = \overline{\left\langle F^{(\alpha)}(t')F^{(\alpha)*}(t+t')\right\rangle} \tag{6.23}$$

(2) If the spin pairs are uncoupled then it is possible to write the relaxation rate taking into account all dipoles formed with a reference spin. The spin–lattice relaxation is still mono-exponential but the relaxation rate is given by

$$\frac{1}{T_1} = \frac{3}{2}K_{dd}^2 I(I+1)\sum_j\left[J_j^{(1)}(\omega_0) + J_j^{(2)}(2\omega_0)\right] \tag{6.24}$$

The density matrix formalism can also be used to obtain the relaxation rates for the spin–spin (e.g., transverse) relaxation rate T_2^{-1} and the spin–lattice relaxation rate in the rotating frame $T_{1\rho}^{-1}$.

For the transverse relaxation due to the direct dipolar coupling of a pair of proton spins the relaxation rate is given by (Abragam, 1961; Kimmich, 1997)

$$\frac{1}{T_2} = \frac{3}{2}K_{dd}^2 I(I+1)\left[\frac{1}{4}J^{(0)}(0) + \frac{5}{2}J^{(1)}(\omega_0) + \frac{1}{4}J^{(2)}(2\omega_0)\right] \tag{6.25}$$

It also possible to obtain the spin–lattice relaxation rate in the rotating frame, $T_{1\rho}^{-1}$, when $\omega_0 \gg \omega_1 \gg \omega_{\text{loc}}$ (Abragam, 1961; Kimmich, 1997)

$$\frac{1}{T_{1\rho}} = \frac{3}{2}K_{dd}^2 I(I+1)\left[\frac{1}{4}J^{(0)}(2\omega_1) + \frac{5}{2}J^{(1)}(\omega_0) + \frac{1}{4}J^{(2)}(2\omega_0)\right] \tag{6.26}$$

6.2.1 Spin–Lattice Relaxation in Aligned Systems

In the case of samples where it is possible to define an alignment direction of molecules (e.g., liquid crystals), it is necessary to consider the angle Δ between the alignment direction and the magnetic field. The spectral densities have to be calculated taking into account the existence of this angle between the two-axis frame of references (Ukleja et al., 1976)

$$J^{(k)}(\omega, \Delta) = \sum_{n=0}^{2} f_{kn}(\sin\Delta)\, J^{(n)}(\omega), \tag{6.27}$$

where

$$f_{kn}(x) = \frac{1}{8}\begin{bmatrix} 8-24x^2+18x^4 & 144(x^2-x^4) & 9x^4 \\ 2(x^2-x^4) & 8-20x^2+16x^4 & 2x^2-x^4 \\ 2x^4 & 16(2x^2-x^4) & 8-8x^2+x^4 \end{bmatrix}. \tag{6.28}$$

Two limit cases are worth mentioning. The isotropic case where the spectral densities present the relative proportions

$$J^{(k)}(\omega) = c_k J^{(1)}(\omega) \tag{6.29}$$

with $c_0 = 6, c_1 = 1$ e $c_2 = 4$. In this case

$$\begin{aligned} J^{(k)}(\omega, \Delta) &= \sum_{n=0}^{2} f_{kn}(\sin\Delta)\, c_n\, J^{(1)}(\omega) \\ &= c_k\, J^{(1)}(\omega) \\ &= J^{(k)}(\omega) \end{aligned} \tag{6.30}$$

and the spectral densities are obviously independent of Δ.

The second case is that of a sample with domains presenting an uniform distribution of orientation angles (e.g., like in a polycrystal). In this case it is necessary to average Eq. 6.28

$$\overline{f_{kn}} = \frac{1}{4\pi}\int_0^{2\pi}\int_0^{\pi} f_{kn}(\sin\Delta)\sin\Delta d\varphi d\Delta = \begin{bmatrix} \frac{2}{5} & \frac{24}{5} & \frac{6}{5} \\ \frac{1}{15} & \frac{4}{5} & \frac{1}{5} \\ \frac{4}{15} & \frac{16}{15} & \frac{4}{5} \end{bmatrix}. \tag{6.31}$$

Equation 6.27 becomes independent of Δ

$$J^{(k)}(\omega) = \sum_{n=0}^{2} \overline{f_{kn}}\, J^{(n)}(\omega). \tag{6.32}$$

6.3 Relaxation Mechanisms

6.3.1 Isotropic Rotations

To be able to calculate the spin–lattice relaxation rate for the nuclear spins systems of a sample it is necessary to consider the motions of the nuclear spins. As they are at the nucleus of atoms their motions are the motions of those atoms in the molecules. In the following we will consider the ^{1}H spins with $I = 1/2$. These motions can be divided in two groups: *internal* motions and *global* motions. The former are the motions of hydrogen atoms inside the molecule associated with fast conformational changes involving chemical groups (e.g., CH_2, CH_3) in aliphatic chains, or rotations of phenyl rings around the para-axis, etc. The latter involve the motions of the molecules as a whole as can translational displacements and rotations/reorientations. In partially ordered systems like liquid-crystalline (LC) phases slow motions involving groups of molecules are identified as *collective* motions.

The calculus of a spin lattice relaxation can be illustrated by considering the rotations of two spins in a "spherical" molecule of radius a. To calculate Eq. 6.20 it is necessary to calculate the probability function $P(\Omega, \Omega_0, t)$ that relates the angular orientation of the interproton spin vector $\Omega_0 = (\theta_0, \varphi_0)$ at time t_0 with the angular orientation $\Omega = (\theta, \varphi)$ at time $t_0 + t$.

$$\frac{\partial P}{\partial t} = \frac{D_R}{a^2}\nabla^2 P \tag{6.33}$$

where D_R is the rotational diffusion constant. A classical solution for this equation involves the expression of the probability function in an expansion of spherical harmonics $P(\Omega, t) = \sum_{m,l} c_l^m(t) Y_l^{(m)}(\Omega)$, where the time dependent functions $c_l^m(t)$ can be obtained from Eq. 6.33 using the know relation $\nabla^2 Y_l^{(m)}(\Omega) = -l(l+1)Y_l^{(m)}(\Omega)$

$$c_l^m(t) = c_l^m(0)e^{-t/\tau_l} \tag{6.34}$$

with $\tau_l^{-1} = D_R^2 l(l+1)a^{-2}$.

The calculus of Eq. 6.20 can performed taking advantage of the orthogonal properties of the spherical harmonics which yields (Abragam, 1961)

$$G^{(1)}(t) = \frac{2}{15r^6}e^{-|t|/\tau_2}, \quad G^{(0)} = 6G^{(1)}, \quad G^{(2)} = 4G^{(1)} \tag{6.35}$$

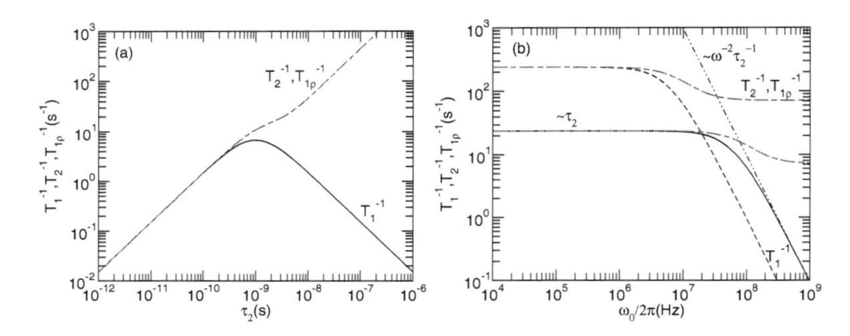

Figure 6.2 Proton relaxation rates calculated with Eq. 6.37, considering $r = 2 \times 10^{-10}$ m and $\nu_1 = \omega_1/(2\pi) = 10$ kHz. (a) The relaxation rates are represented as functions of τ_2 for $\nu_0 = \omega_o/(2\pi) = 100$ MHz. (b) The relaxation rates are represented as functions of the Larmor frequency for $\tau_2 = 1/(2\pi 10^8)$ s (black solid line) and $\tau_2 = 1/(2\pi 10^7)$ s (black dashed line). Spin–spin relaxation rate and spin–lattice relaxation in the rotating frame calculated using Eqs. 6.38 and 6.39, respectively, are also presented.

with $\tau_2 = D_R/(6a^2)$. The calculus of the spectral densities is straightforward and gives

$$J^{(1)}(\omega_0) = \frac{4}{15r^6}\frac{\tau_2}{1+\omega_o^2\tau_2^2}, \quad J^{(0)} = 6J^{(1)}, \quad J^{(2)} = 4J^{(1)} \tag{6.36}$$

The spin–lattice relaxation rate is

$$\frac{1}{T_1} = \frac{3}{10}\frac{K_{dd}^2}{r^6}\left[\frac{\tau_2}{1+\omega_o^2\tau_2^2} + \frac{4\tau_2}{1+4\omega_o^2\tau_2^2}\right] \tag{6.37}$$

and the spin–spin relaxation rate is

$$\frac{1}{T_2} = \frac{3}{10}\frac{K_{dd}^2}{r^6}\left[\frac{3}{2}\tau_2 + \frac{5}{2}\frac{\tau_2}{1+\omega_o^2\tau_2^2} + \frac{\tau_2}{1+4\omega_o^2\tau_2^2}\right] \tag{6.38}$$

The spin–lattice relaxation rate in the rotating frame is

$$\frac{1}{T_{1\rho}} = \frac{3}{10}\frac{K_{dd}^2}{r^6}\left[\frac{3}{2}\frac{\tau_2}{1+\omega_1^2\tau_2^2} + \frac{5}{2}\frac{\tau_2}{1+\omega_o^2\tau_2^2} + \frac{\tau_2}{1+4\omega_o^2\tau_2^2}\right] \tag{6.39}$$

In Fig. 6.2 are presented the relaxation rates calculates using Eqs. 6.37, 6.38, and 6.39. $K_{dd} \simeq 8 \times 10^{-25}$ m^{3}s^{-1}.[a] $r = 2 \times 10^{-10}$ m

[a]From Eq. 6.9 using $\gamma_I = \gamma_S = 2.675\ 221\ 28(81) \times 10^8$ s^{-1}T^{-1}, $\mu_0 = 4\pi \times 10^{-7}$ Hm^{-1}, $\hbar = h/(2\pi)) = 1.054\ 571\ 628(53) \times 10^{-34}$ Js, $K_{dd} \simeq 8 \times 10^{-25}$ m^{3}s^{-1} (IUPAC, 1997).

and $\nu_1 = \omega_1/(2\pi) = 10$ kHz. As it can be observed in Fig. 6.2a the longitudinal relaxation rate is represented as a function of τ_2 and shows a maximum for a value of $\tau_2 \approx \omega_o^{-1}$. Since $\omega_1\tau_2 \ll 1$, T_2^{-1} and $T_{1\rho}^{-1}$ are very similar within the plot's resolution and show an inflexion around $\omega_0\tau_2 \approx 1$ and are proportional to τ_2 when $\omega_0\tau_2 \gg 1$. In Fig. 6.2b are presented the relaxation rates calculated as functions of the Larmor frequency, for two values of τ_2. In both cases, it is clear that in the frequency range $\omega_0\tau_2 \ll 1$ all relaxation rates have similar values and for values of τ_2 and ω_1 that verify $\omega_1\tau_2 \ll 1$ $T_1^{-1} \approx T_2^{-1} \approx T_{1\rho}^{-1} \approx 5\tau_2$. In the high-frequency limit $\omega_0\tau_2 \gg 1$ $T_1^{-1} \approx \omega_0^{-2}$ whereas $T_2^{-1} \approx T_{1\rho}^{-1} \approx 3\tau_2/2$ became independent of the Larmor frequency.

In Fig. 6.2b it is possible to observe that an increase of τ_2 produces an increase of the relaxation rates for frequencies $\omega_0 \ll \tau_2^{-1}$. For frequencies $\omega_0 \gg \tau_2^{-1}$, the spin–lattice relaxation rate has the opposite behavior as it decreases with increasing τ_2. In the high-frequency regime both T_2^{-1} and $T_{1\rho}^{-1}$ increase with τ_2.

This type of analysis becomes handy when considering the temperature dependence of correlation time τ_2. The molecular rotations are usually thermally activated processes and the rotation diffusion is usually expressed by an Arrhenius temperature dependence expression

$$D_R = D_{R\infty}e^{-E_a/(RT)} \tag{6.40}$$

with $R = N_A k_B$, where N_A is the Avogadro's number[a] and k_B is the Boltzmann constant.[b] E_a is usually referred to as *activation energy* and in some cases of the order of kJ per mol. Since $\tau_2 \propto D_R$ it is clear that when the temperature increases the correlation time decreases (i.e., the motions are faster).

Therefore, with respect to molecular rotations when studied by proton spin–lattice relaxation as described by Eq. 6.37, it is possible to relate an increase of temperature to a decrease of the relaxation rate at low frequencies ($\omega_0\tau_2 \ll 1$) and to an increase of T_1^{-1} at high frequencies ($\omega_0\tau_2 \gg 1$).

In the case of more than one pair of identical spins when the rotations of the molecules can be still described by a single

[a] Avogadro's number: $N_A = 6.022\ 141\ 79(30) \times 10^{23}$ mol^{-1} (IUPAC, 1997).
[b] $k_B = 1.380\ 650\ 4(24)^{-23}$ JK^{-1} (IUPAC, 1997).

correlation time to fast spin–spin interactions Eqs. 6.37, 6.38, and 6.39 are still valid provided that and ensemble average is performed over all interspin vectors, leading to spectral densities of the form

$$J^{(1)}(\omega_0) = \frac{4}{15}\left\langle \frac{1}{r^6} \right\rangle \frac{\tau}{1+\omega_0^2\tau^2}. \tag{6.41}$$

The relaxation mechanism expressed by Eq. 6.37 with the spectral densities given by Eq. 6.36 is often referred to as the Bloembergen–Purcell–Pound (BPP) relaxation mechanism (Bloembergen et al., 1948).

6.3.2 Rotations/Reorientations

In the case of anisometric molecules (e.g., elongated ellipsoidal molecules or disc-like molecules) both rotations and/or reorientations can coexist and its contribution to modulate the spin–lattice relaxation requires a good description of the rotational diffusion process taking into account the anisotropy of the local environment. In the case of liquid crystals molecular order breaks the otherwise rotations isotropy and introduces a degree of complexity in the relaxation model. Different models have been proposed to describe rotations in these systems (Dong, 1997; Sebastião et al., 2009).

One model often used is the extended Woessner model for the relaxation induced by reorientations of rigid elongated molecules (Beckmann et al., 1983, 1986; Dolinšek et al., 1991; Dong, 1997; Rutar et al., 1978; Sebastião et al., 1992; Vilfan et al., 2007; Woessner, 1962). The spectral densities are given by:

$$J_k(\omega_0) = \frac{4}{3}c_k \sum_{m=-2}^{2} \frac{\overline{|d_{k0}^2(\alpha_{ij})|^2}}{a_{ij}^6} \left\langle |D_{km}^2(\theta)|^2 \right\rangle \frac{\tau_{|m|}}{1+\omega_0^2\tau_{|m|}^2} \tag{6.42}$$

where $c_0 = 6$, $c_1 = 1$, and $c_2 = 4$. a_{ij} denotes the distance between the proton pair and α_{ij} the angle between the interproton vector and the long molecular axis. D_{km}^2 is the second rank Wigner rotation matrix and the averages $\left\langle |D_{km}^2(\theta)|^2 \right\rangle$ can be expressed in terms of the second and fourth rank Legendre polynomials of the angle θ between the long molecular axis and the nematic director, that is, the nematic order parameter S and $\langle P_4 \rangle$ (Dolinšek et al., 1991; Dong, 1997), respectively.

$$\left\langle |D_{k,\pm m}|^2 \right\rangle =$$

$$\begin{bmatrix} 7+10S+18\langle P_4\rangle-35S^2 & 7+5S-12\langle P_4\rangle & 7-10S+3\langle P_4\rangle \\ 7+5S-12\langle P_4\rangle & 7+\frac{5}{2}S+8\langle P_4\rangle & 7-5S-2\langle P_4\rangle \\ 7-10S+3\langle P_4\rangle & 7-5S-2\langle P_4\rangle & 7+10S+\frac{1}{2}\langle P_4\rangle \end{bmatrix} \tag{6.43}$$

with

$$\begin{aligned} S = \langle P_2 \rangle &= \frac{1}{2}\left\langle 3\cos^2\theta - 1 \right\rangle \\ \langle P_4 \rangle &= \frac{1}{8}\left\langle 35\cos^4\theta - 30\cos^2\theta + 3 \right\rangle . \end{aligned} \tag{6.44}$$

The higher-order parameter $\langle P_4 \rangle$ can be to a good approximation expressed as $5/7S^2$ (Fabbri and Zannoni, 1986).

$d^2_{km}(\alpha_{ij})$ is the second rank reduced Wigner rotation matrix and the expressions for the factors $\overline{|d^2_{k0}(\alpha_{ij})|^2/a^6_{ij}}$ are (Dolinšek et al., 1991; Sebastião et al., 1992):

$$\overline{|d^2_{k0}(\alpha_{ij})|^2/a^6_{ij}} = \begin{cases} \overline{(3\cos^2\alpha_{ij}-1)^2/(4a^6_{ij})}, & k=0 \\ \overline{3\sin^2 2\alpha_{ij}/(8a^6_{ij})}, & k=1 \\ \overline{3\sin^4\alpha_{ij}/(8a^6_{ij})}, & k=2 \end{cases} \tag{6.45}$$

The correlation times $\tau_{|m|}$ can be expressed in terms of two correlation times corresponding to molecular reorientations with respect to directions parallel and perpendicular to the long molecular axis, τ_L and τ_S, respectively (Dolinšek et al., 1991; Vilfan et al., 2007):

$$\begin{aligned} \tau_0^{-1} = \tau_S^{-1},\ \tau_1^{-1} &= \frac{1}{\tau_L} + \frac{1}{\tau_S}, \\ \tau_2^{-1} &= \frac{4}{\tau_L} + \frac{1}{\tau_S}. \end{aligned} \tag{6.46}$$

The original Woessner's expression for the isotropic phase (Woessner, 1962) is obtained by inserting $S = 0$ and $\langle P_4 \rangle = 0$ into Eq. 6.43.

Specific models can be found in the literature to describe the spin–lattice relaxation mechanism associated with rotations/reorientations of different type of calamitic molecules for different reorientation potentials (Dong, 1997). In general terms these models are expressed by sets of correlation times in an attempt to detail the rotational diffusion motions of the molecules or molecular segments.

Due to the fact that conformational changes have correlation times typically of the order of $\sim 10^{-12}$ s, these internal spin motions are only accessible by T_1^{-1} measurements at frequencies above hundreds of megahertz.

In the case of dendrimers, the description of molecular rotations reorientations has to take into account the particular molecular structure and the fact that molecular interdigitation might hinder global molecular rotations/reorientations. One way to approach this problem is to assume different correlation times for particular molecular segments and/or groups, like the internal dendrimer core molecular chains and groups and the peripheral mesogenic groups. Rotations/reorientations in these molecular systems can be described by a sum of BPP (Eq. 6.37) contributions.

6.3.3 Molecular Translational Self-Diffusion

Molecular translational self-diffusion (SD) can be detected in many systems with different degrees of importance. Proton NMR relaxometry can detect translational molecular displacements since these motions modulate intermolecular proton spin interactions. Depending on the viscosity of the medium the relaxation contribution for this type of molecular motion can be more effective than other relaxation contributions in different frequency ranges. In the case of dendrimers their high viscosity might strongly reduce the importance of the SD of the whole dendrimer molecules. However, the motion of the peripheral mesogenic units might undergo displacements due to the dendrimers' internal core conformation changes that might be understood as a type of restricted diffusion process. To the present no specific relaxation model was developed to quantify the contribution to relaxation associated to this particular motions.

In the case of individual anisometric molecules undergoing translational displacements the spin–lattice relaxation can be expressed in terms of spectral densities of the type

$$J^{(k)}(\omega_0)_{\mathrm{SD}} = \frac{n\tau_{\mathrm{D}}}{d^3} j_{\mathrm{SD}}^{(k)}(\omega_0 \tau_{\mathrm{D}}) \tag{6.47}$$

where n is the density of spins, τ_{D} is the mean jump time, d is the distance of closest approach between molecules, and $j_{\mathrm{SD}}^{(k)}(\omega_0 \tau_{\mathrm{D}})$ are dimensionless functions that depend on the phase structure, for example, isotropic (Harmon and Muller, 1969; Torrey, 1953) or LC phase (Blinc et al., 1974; Žumer and Vilfan, 1978, 1980, 1981, 1983).

In the case of small molecules confined in nanoscale cavities, molecular translational (SD) displacements can induce rotations/reorientations of the molecules due to the topology of the cavities and surface walls. In these systems it is necessary to consider a specific relaxation mechanism that considers the effect of reorientations mediated by translational diffusion (RMTD) (Kimmich and Weber, 1993; Stapf et al., 1994; Zavada and Kimmich, 1999, 1998). The spectral densities obtained for this relaxation mechanism produce relaxation dispersion profiles that differ considerably from those related SD in bulk. For the case of an isotropic distribution of cavities' orientations, the spectral densities are written as

$$J_{\mathrm{RMTD}}^{(k)}(\omega) = c_k \int_{q_{\min}}^{q_{\max}} Q(q) \frac{2\tau_q}{1+\omega^2\tau_q^2} dq \tag{6.48}$$

with c_k defined as for Eq. 6.42. $Q(q)$ is the orientational structure factor function which depends on wave numbers q of diffusive modes that decay exponentially with characteristic times τ_q. Depending on the structure factor $Q(q)$, different frequency dependencies can be observed for the spin–lattice relaxation rate typically expressed by power laws $T_1^{-1} \approx \nu^{-p}$ (Sebastião et al., 2005; Vilfan et al., 2007).

6.3.3.1 Collective motions

Liquid crystal phases present specific structural properties that allow for elastic deformations that involve collective motions of a large number of molecules. For the most common LC mesophases the collective motions can be associated to *order director fluctuations* (ODFs) in nematic phases, *layer undulations* in liquid-like layered phases, and *column undulations* in columnar phases.

The relaxation models associated to the above collective motions take into account the complex viscoelastic properties of the systems. In the case of the nematic phase the elastic free energy is given by the previously introduced Frank's equation (Eq. 2.7) that presents three elastic constants K_1, K_2, and K_3 for the three distortion modes of the nematic director field, splay, twist, and bend, respectively. Also, the viscosity in the nematic phase is described by five independent coefficients (de Gennes and Prost, 1993), so-called *Leslie coefficients*.

The Gibbs free-energy expression for the elastic deformations in nematics can be written in terms of a wave vector in a conjugated space by a Fourier transform of $\vec{n}(\vec{r})$

$$\vec{n}(\vec{q}) = \int \vec{n}(\vec{r}) e^{i\vec{q}.\vec{r}} d\vec{r}. \tag{6.49}$$

Then in the Frank energy expression, Eq. 2.7, each nematic director field component (n_x, n_y, n_z) can be expanded in Fourier series

$$n_\alpha(\vec{r}) = \frac{1}{V} \sum_{\vec{q}} n_\alpha(\vec{q}) e^{-i\vec{q}.\vec{r}}. \tag{6.50}$$

With this expansion the Frank energy can be further diagonalized by transforming n_x and n_y in n_1 and n_2, respectively, where n_1 describes periodic distortions involving splay and bend deformations and n_2 describes periodic distortions involving twist and bend. By defining $q_\parallel^2 = q_z^2$ and $q_\perp^2 = q_x^2 + q_y^2$ it is possible to write

$$F = \frac{1}{2V} \sum_{\vec{q}} \sum_{\alpha}^{2} \mathcal{K}_\alpha(\vec{q}) |n_\alpha(\vec{q})|^2, \tag{6.51}$$

where the functions $\mathcal{K}_\alpha(\vec{q})$ are specific for each mesophase, as they depend on the model for the elastic deformations. For nematic and smectic A phases, neglecting the compression of the layers in the latter,

$$\mathcal{K}_\alpha(\vec{q}) = K_\alpha q_\perp^2 + K_3 q_\parallel^2. \tag{6.52}$$

The deformation modes $n_\alpha(\vec{q})$ that reflect the return to the equilibrium of the director field have the an exponential decaying time dependence (de Gennes and Prost, 1993)

$$\frac{\partial n_\alpha(\vec{q})}{\partial t} = \frac{1}{\tau_\alpha(\vec{q})} n_\alpha(\vec{q}) \tag{6.53}$$

with time constants

$$\tau_\alpha(\vec{q}) = \frac{\eta_\alpha(\vec{q})}{\mathcal{K}_\alpha(\vec{q})}. \tag{6.54}$$

The spectral densities can be calculated taking into account the autocorrelation between the distortions in the nematic field as a function of position and time

$$\langle \delta n(\vec{r}, 0)\delta n(\vec{r}, t)\rangle\,. \tag{6.55}$$

The calculation of the spectral densities is made performing space and time Fourier transforms to a fluctuation modes space $\vec{q}$ and to the frequency domain.

First,

$$\langle \delta n(\vec{q}, 0)\delta n(\vec{q}, t)\rangle = \left\langle n^2(\vec{q})\right\rangle e^{-t/\tau(\vec{q})} \tag{6.56}$$

with

$$\left\langle n^2(\vec{q})\right\rangle = \frac{k_B T}{V\mathcal{K}_\alpha(\vec{q})}. \tag{6.57}$$

In general terms the spectral densities can be expressed as

$$\begin{aligned} J^{(k)}(\nu) \;=\; & f_{k1}(\Delta)\, S^2\, \frac{\overline{|d^2_{00}(\alpha_{ij})|^2}}{r^6_{ij}}\, \frac{k_\mathrm{B} T}{(2\pi)^3}\,\cdot \\ & \cdot \sum_{\alpha=1}^{2} \iiint d^3q\, \frac{1}{\mathcal{K}_\alpha(\vec{q})}\, \frac{2\tau_\alpha(\vec{q})}{1 + 4\pi^2\nu^2\tau^2_\alpha(\vec{q})}. \end{aligned} \tag{6.58}$$

In the case of the N phase, as a first and general approximation the elastic constants and the viscosity parameters can be reduce to one elastic constant, K, and one effective viscosity, η. Therefore,

$$\mathcal{K}_\alpha(\vec{q}) = Kq^2 \tag{6.59}$$

and the spectral densities for the ODF contribution to the relaxation rate may be written as (Blinc et al., 1975; Doane et al., 1974; Freed, 1977; Vold and Vold, 1994; Zupančič et al., 1976)

$$\begin{aligned} J^{(k)}_\mathrm{ODF}(\omega, \Delta) \;=\; & f_{k1}(\Delta)\, \frac{\overline{|d^2_{00}(\alpha_{ij})|^2}}{a^6_{ij}}\, \frac{k_\mathrm{B} T\, S^2 \eta^{1/2}}{\pi\sqrt{2}\, K^{3/2}}\, \omega^{-1/2}\,. \\ & \left[f\left(\frac{\omega_\mathrm{cM}}{\omega}\right) - f\left(\frac{\omega_\mathrm{cm}}{\omega}\right)\right] \end{aligned} \tag{6.60}$$

where

$$f(x) = \frac{1}{\pi}\left[\tan^{-1}(\sqrt{2x}+1) + \tan^{-1}(\sqrt{2x}-1) - \tanh^{-1}\frac{\sqrt{2x}}{1+x}\right] \quad (6.61)$$

T is temperature, η is the effective viscosity, K the elastic constant (in a one constant approximation), k_B is the Boltzmann constant, and $f(x)$ is the cutoff function. $\omega_{cM} = Kq_{max}^2/\eta$ and $\omega_{cm} = Kq_{min}^2/\eta$ are the high and low cutoff frequencies, respectively. q_{max} and q_{min} are the highest and the lowest wave number of the fluctuation modes, respectively. The $T_1^{-1} \propto \omega^{-1/2}$ law that is obtained from Eqs. 6.60 and 6.61 for the ODF relaxation mechanism in N phases was first introduced by Pincus (Pincus, 1969). Translational SD can contribute to the damping of the fluctuation modes. When SD is not considered as an independent relaxation mechanism, this effect can be included in the spectral densities by modifying the ratio $K/\eta \rightarrow K/\eta + D$ (Dong, 1997; Lubensky, 1970; Pincus, 1969).

In the SmA and lamellar phases of lyotropic systems, the collective motions associated with director fluctuations give a contribution from molecular displacements that preserve the layer's structure—*layer undulations*. In this case it is generally assumed that the only deformation that prevails is splay and the only Frank elastic constant to take into account is

$$\mathcal{K}_\alpha = K_1 q_\parallel^2. \quad (6.62)$$

In this case the spectral densities are given by (Blinc et al., 1975; Carvalho et al., 2001a; Sebastião et al., 1993, 1995; Vilfan et al., 2007)

$$J_k^{LU}(\omega,\Delta) = f_{k1}(\Delta)\frac{\overline{|d_{00}^2(\alpha_{ij})|^2}}{a_{ij}^6}\frac{k_B T S^2}{2K_1 L}\omega^{-1}\frac{1}{\pi} \times\left[\tan^{-1}\left(\frac{\omega_{cM}}{\omega}\right) - \tan^{-1}\left(\frac{\omega_{cm}}{\omega}\right)\right], \quad (6.63)$$

where K_1 is the splay elastic constant, and L is the correlation length of undulations in the direction perpendicular to the layers. For $\omega_{cM} \ll \omega \ll \omega_{cm}$ the relaxation rate is $\left(T_1^{-1}\right)_{LU}$ is proportional to ω^{-1}.

In the case of the smectic C phase, in-plane fluctuations of the tilt angle direction can occur. These fluctuations have a nematic-like character q wave vector fluctuations in all directions can occur.

This contribution is proportional to the inverse square root of the Larmor frequency, as for the N case $\left(T_1^{-1}\right)_{DF_N} \propto \omega^{-1/2}$ (Carvalho et al., 2001b).

In columnar phases, the collective motions are associated to columns' undulations expressed by a relaxation mechanism described by the Žumer and Vilfan's model with

$$\mathcal{K}_\alpha = K_3 q_\perp^2 + B\frac{q_\perp^2}{q_\parallel^2} \tag{6.64}$$

and

$$J_{ECD}^{(k)}(\omega_0, \Delta) = f_{k1}(\Delta) C\,\omega^{-1} \int_0^1 \left[\tan^{-1}\left(\frac{R\omega_c}{u\omega} + u\frac{\omega_c}{\omega}\right) - \tan^{-1}\left(u\frac{\omega_c}{\omega}\right)\right] du, \tag{6.65}$$

where C is a constant that depends on the viscoelastic properties in the columnar phase of the LCs, namely K_3 and B, R depends on the elastic constants corresponding to the bending and compression of the columns and ω_c is a cutoff frequency (Cruz et al., 1996, 1998; Žumer and Vilfan, 1981) . Equation 6.65 is obtained using a single constant approximation with respect to the deformation of the 2D lattice in the plane perpendicular to the columns.

6.3.3.2 Cross-relaxation

In the case of multiple-spin species present in the system there is the possibility to observe cross-relaxation between the hydrogen spin population and other spin systems. Although in most cases molecules are composed of elements with different nuclear spins due to their relative abundance and/or weak coupling between the different spins subsystems and/or the difference between their Larmor frequencies, there are not many cases where this cross-relaxation is observed (Gradišek et al., 2014). Another type of cross-relaxation can be observed between the hydrogen spins' system and other spin system for which the main relaxation is due to quadrupole interactions. This type of cross-relaxation is more common and has been reported for materials with molecules possessing nitrogen and/or chlorine atoms and/or deuterium, all with nuclear spins larger than 1/2.

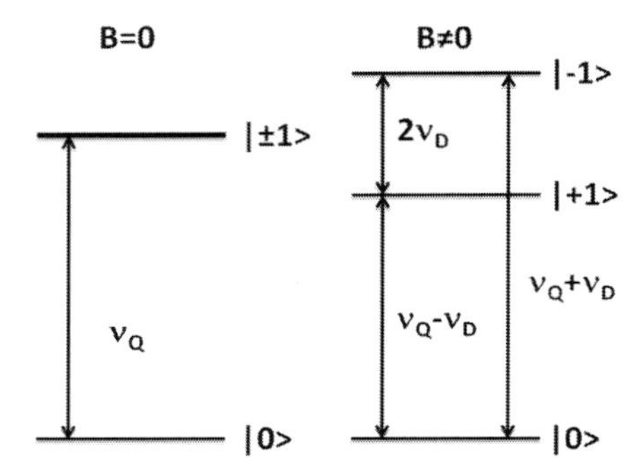

Figure 6.3 Energy levels and eigenstates for a spin $I = 1$ in zero and nonzero magnetic field (original figure from (Gradišek et al., 2014)). Reprinted with permission from Gradišek, A., Sebastião, P. J., Fernandes, S. N., Apih, T., Godinho, M. H. and Seliger, J. (2014). (1)h-(2)h cross-relaxation study in a partially deuterated nematic liquid crystal, *Journal of Physical Chemistry. B* **118**, 20, pp. 5600–5607. Copyright (2014), American Chemical Society.

Since the energy levels of the $I = 1$ spin system depend on the magnetic field as illustrated in Fig. 6.3 it is possible to measure the spin–lattice relaxation of the $I = 1/2$ spin system (ex: ^{1}H system) as a function of the magnetic field (i.e., equivalent to, say, the Larmor frequency, due to $\omega_0 = \gamma B$) and observe that there is a match between the energy levels of both spins systems at certain

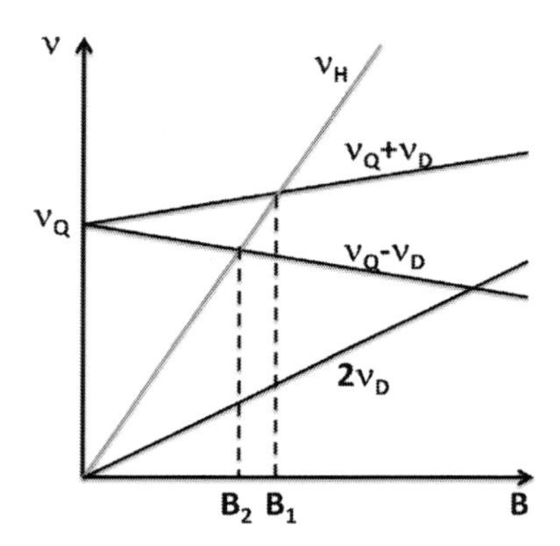

Figure 6.4 Resonance frequencies in dependence on the magnetic field **B** for spin $I = 1/2$ and spin $I = 1$. The resonance frequencies match at $\mathbf{B} = \mathbf{B}_1$ and $\mathbf{B} = \mathbf{B}_2$.

frequencies, as illustrated in Fig. 6.4. Therefore, for frequencies in the vicinity of cross-relaxation frequencies $\omega_{CR1} = 2\pi\nu_Q + \gamma_{I=1}B$ and $\omega_{CR2} = 2\pi\nu_Q - \gamma_{I=1}B$ the relaxation rate of the $I = 1/2$ system will have an additional relaxation mechanism through the quadrupolar spins' system. Usually this observed as a "cusp like" increase of the relaxation rate at two frequencies.

Lorentzian curves centered at frequencies ω_{CR1} and ω_{CR12} are commonly used to model these additional relaxation contributions

$$\left(\frac{1}{T_1}\right)_{CR} = \sum_i A_i \frac{\tau_i}{1 + (\omega - \omega_i)^2 \tau_i^2}. \tag{6.66}$$

These relaxation contributions are effective mainly at in the frequency regions localized around the quadrupolar resonance frequencies ω_i and are observed in compounds with elements with nuclear spins larger than 1/2. In the following cross-relaxation contributions were detected in dendrimer systems where the mesogenic terminal groups possess nitrogen atoms.

Chapter 7

NMR Relaxometry and Molecular Dynamics: Experimental Techniques

The measurement of nuclear magnetic resonance (NMR) relaxation times T_1, T_2, and $T_{1\rho}$ is possible using appropriate experimental equipment and suitable experimental conditions. Depending on each particular relaxation time different experimental techniques have to be used. Here is presented a brief review of the experimental setups and techniques most often used.

Due to the fact that the NMR signal detected has a signal-to-noise ratio that decreases with $B_0^{3/2}$ some experimental techniques can be applied in practice only for $B_0 > 0.1$ T (e.g., ^{1}H-NMR frequency larger than 4 MHz). Above this magnetic field many radio-frequency (RF) pulse sequences can be used to measure the relaxation times T_1, T_2, and $T_{1\rho}$.

7.1 Inversion Recovery

For T_1 the *inversion recovery* pulse sequence is the one mostly used. This RF pulse sequence is composed of a π-pulse, followed by a relaxation time τ and a $\pi/2$-pulse. The FID signal is collected after

NMR of Liquid Crystal Dendrimers
Carlos R. Cruz, João L. Figueirinhas, and Pedro J. Sebastião

ISBN 978-981-4745-72-7 (Hardcover), 978-981-4745-73-4 (eBook)
www.panstanford.com

the second pulse. A complete spin–lattice relaxation is required before repeating the pulse sequence, which requires a time $\tau_P \gg 5T_1$.

$$\pi \to \tau \to \pi/2 \to \mathrm{FID} \to (\sim 5 \times T_1) \to \ldots . \quad (7.1)$$

The amplitude of the RF field B_1 and the duration of the RF pulse are adjusted so that the effect upon the magnetization of the spins' system can be regarded, in a classical picture, as the rotation of the magnetization vector in the laboratory frame in a way that the angles defined by the initial (before the RF pulse) orientations have determined angles (ex: π, $\pi/2$), as referred to in Chapter 5. The pulse duration is thus obtained from the relation $\theta = \gamma B_1 t_\theta$.

In a classical picture, the π-pulse flips the magnetization from its initial equilibrium state, aligned with the external magnetic field, $\overrightarrow{M}(0) = M_0\vec{e}_z \parallel B_0\vec{e}_z$, to an antiparallel state $\overrightarrow{M}(0 + t_\pi) = -M_0\vec{e}_z$. The $\pi/2$ pulse transfers the magnetization to the plane $xy \perp \vec{e}_z$ and allows for the detection of the free induction decay signal. During time τ the magnetization component M_z evolves according to the solutions of Bloch's equations (Eq. 4.22)

$$M_z = M_0\left[1 - 2\exp\left(-\tau/T_1\right)\right]. \quad (7.2)$$

Depending on the acquisition system it is possible to include phase cycles in the relative phases of the RF field so that the both real and imaginary parts of the FID response and both positive and negative signals can be obtained in order to minimize and/or compensate for DC offsets.

7.2 Fast Field Cycling

In the case of Zeeman fields that correspond to Larmor frequencies lower than 4 MHz fast-field-cycling (FFC) techniques can be used to measure T_1 avoiding the low signal-to-noise ratio that would be found if the conventional inversion recovery sequence would be used.

In a typical FFC experiment the Zeeman field is cycled between different values. Figure 7.1 presents a schematic of a time diagram of a basic field cycle. In this figure the Zeeman magnetic field

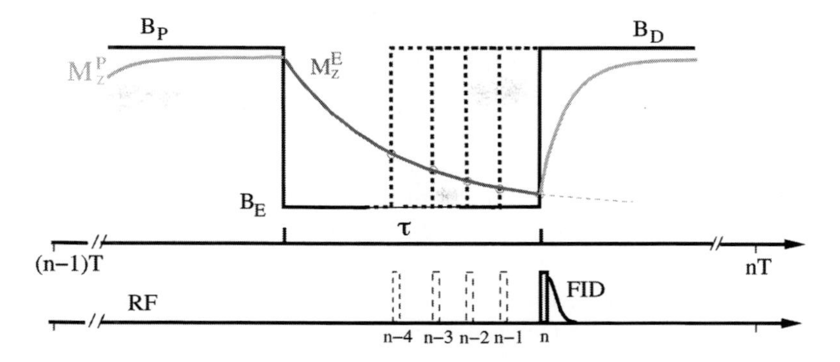

Figure 7.1 Zeeman field cycle in an FFC experiment.

is kept always parallel to the z direction and its amplitude is cycled up and down between three levels with values B_P, B_E, and B_D, with P, E, and D standing for *polarization*, *evolution*, and *detection*, respectively.[a] Due to this magnetic field transitions the magnetization component M_z evolves with time according the Bloch's equation (Eq. 4.18) in the absence of RF fields. It easy to understand that after time τ (see Fig. 7.1) the magnetization is at the lowest value and that the evolution back to the initial state will retain the information about $M_z^E(\tau)$. Since the RF pulse is applied when the sample is submitted to the Zeeman field B_D and the FID's signal-to-noise ratio, that depends on the value of $B_D^{3/2}$ is larger than the one that would be obtained if the FID had been obtained with an RF in resonance with the Larmor frequency $\omega_E = \gamma B_E$. Since the FID detected has an initial amplitude proportional to the value of $M_z^E(\tau)$ the measure of $T_1(B_E)$ can be performed by repeating the previous sequence for different values of τ.

A suitable selection of values τ will allow for the sampling of M_z^E as a function of τ. Since during the time τ the sample is submitted to the magnetic field B_E the evolution of $M_z^E(t)$ depends on the value $T_1^E \equiv T_1(B_E) \equiv T_1(\omega_E)$, with $\omega_E = \gamma B_E$.

The solution of the Block equation for the time interval $[0; \tau]$ is

$$M_z^E(\tau) = M_{z\,\mathrm{eq}}^P - \left(M_{z\,\mathrm{eq}}^P - M_{z\,\mathrm{eq}}^E\right)\left[1 - e^{-\tau/T_1(\omega_E)}\right] \tag{7.3}$$

[a]In a large number of situations $B_P = B_D$.

where, $M_{z\,\mathrm{eq}}^{E} \equiv M_z^E(\tau \to \infty)$ and $M_{z\,\mathrm{eq}}^{P} \equiv M_z^P(\tau_P \to \infty)$. Since the values of B_E extend from B_P to zero is it possible, in principle to measure T_1 for any field below B_P, that is, for large Larmor frequency range, that typically extends from 10 MHz to 5 kHz. The development of new FFC NMR setups is continuously enlarging this frequency range pushing the upper limit to frequencies close to 50 MHz and the lower limit to the earth's field.[a]

Field cycles other than that illustrated in Fig. 7.1 can be used in order to optimize the measuring conditions. In particular, it is necessary to ensure experimental conditions that make possible to determine T_1 with a low uncertainty, typically 5%–10%. In particular a field cycle is required for which $|M_{z\,\mathrm{eq}}^{P} - M_{z\mathrm{eq}}^{E}| > M_{z\,\mathrm{eq}}^{P}/2$, $M_{z\,\mathrm{eq}}^{E}/2$, and eventually the polarization is made for a magnetic $B_P < B_E$. It is also possible measure T_1 for magnetic field values $B_E > B_D$.

It is important to note that the magnetic field transitions $B_P \to B_E$ and $B_E \to B_D$ illustrated in Fig. 7.1 represent the ideal case when $\tau_{\mathrm{off}} \equiv \tau_{B_P \to B_E} \ll T_1(B_E)$ and $\tau_{\mathrm{on}} \equiv \tau_{B_E \to B_D} \ll T_1(B_D)$. In real situations it is necessary to make sure that $\tau_{\mathrm{off}} < T_1(B_E)$, at least. For most FFC equipment the $\tau_{\mathrm{off}} \approx \tau_{\mathrm{on}} \approx$ 1–3 ms. A direct consequence of these finite magnetic field switching times and associated transient effects is that the magnetization decay during τ is actually better described by

$$M_z^E(\tau_{\mathrm{eff}}) \simeq f_{\mathrm{on}}\left[M_{z\,\mathrm{eq}}^{P} + \Delta M_{z\,\mathrm{on}} - \left(M_{z\,\mathrm{eq}}^{P} - M_{z\,\mathrm{eq}}^{E}\right)\left(1 - e^{-\frac{\tau_{\mathrm{eff}}}{T_1(\omega_E)}}\right)\right] \tag{7.4}$$

where $\tau_{\mathrm{eff}} = \tau + \tau_{\mathrm{on}} + \tau_\delta$ represents the effective evolution delay from the $B_P \to B_E$ transition moment to the measuring moment after τ_{on} and stabilization delay τ_δ. f_{on} and $\Delta M_{z\,\mathrm{on}}$ are instrumental factors that associated with the described delays (Noack, 1986).

Another aspect to take into consideration is the fact that τ_{off} and τ_{on} cannot be arbitrarily short since it is necessary to assure that during the magnetic field transitions the nuclear magnetization remains parallel to the external magnetic field. This condition, referred to as *adiabatic switching*, is usually expressed as

$$\left|\vec{B} \times \frac{d\vec{B}}{dt}\right| / B^2 \ll \gamma B. \tag{7.5}$$

[a] T_1 measurements bellow the earth's field are possible if the earth's field is compensated with a suitable setup.

Ideally, $\vec{B} \parallel d\vec{B}/dt$ and the above condition is satisfied. However, besides transverse components that might exist due to technical reasons, nonadiabatic conditions can occur when the value of the external magnetic field is comparable with the local fields $\vec{B}_{\text{loc}}$ present in the material systems studied. In this case the adiabatic condition must be expressed in terms of $\vec{B}_{\text{eff}} = \vec{B} + \vec{B}_{\text{loc}}$ and might be hard to realize. Actually, $\vec{B}_{\text{loc}}$ represents the lowest field achievable since it is intrinsic to the systems and cannot me easily compensated. The external magnetic field $\vec{B}$, on the other hand can be set to zero provided that the earth's magnetic field is properly compensated.

Nonadiabatic switching conditions produce instabilities in T_1 measurement as the NMR signal becomes unreproducible during a sequence of repetitive field cycles. This behavior can be detected usually for magnetic fields below $\sim 4 \times 10^{-4}$ T and depends on the systems studied. One way to avoid the problem is to decrease dB/dt when $B \approx B^E$ by a proper adjustment of the switch down control parameters.

7.3 Experimental Setups

In general terms an experimental NMR setup used in an FFC experiment is illustrated in Fig. 7.2. Actually, the differences with respect to a conventional NMR setup are noticed only in the magnet's power supply. In conventional NMR setups the magnet's power supply (if existing) produces steady currents with the lowest currently ripple possible in order to assure the best field stability possible for a good in-phase signal detection. In FFC power supplies different levels of current with good stability must be generated and fast (0.1–3 ms) switching between levels must be possible.

Besides the magnet's power supply and the magnet a number of auxiliary circuits and systems are included in the setup. For some applications the magnet and its associated power supply can be replaced by a permanent magnet that can be either a superconductor electromagnet or a solid-state magnet. In those setups the magnetic field is static and the spin–lattice relaxation is measured for the particular Larmor frequency corresponding

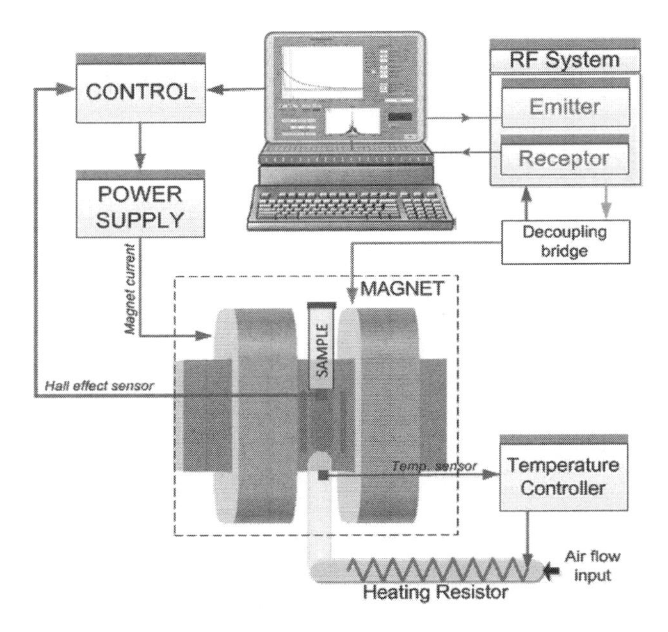

Figure 7.2 Simple block diagram of an FFC relaxometer.

to the static magnetic field. These NMR setups are used for NMR spectroscopy studies or spin–spin relaxation NMR analyses.

These fast current transitions became possible due to advances in power semiconductors. Power supplies with topologies based on bipolar transistors, isolated gate bipolar transistors (IGBTs), and metal oxide semiconductor field-effect transistors (MOSFETs) have been proposed (Kimmich and Anoardo, 2004; Sousa et al., 2004, 2010).

Chapter 8

NMR Spectroscopy of Liquid Crystal Dendrimers: Experimental Results

8.1 NMR of LC Dendrimers and the Investigation of the Biaxial Nematic Phase

8.1.1 Biaxial Nematic Ordering and NMR Spectroscopy

As described in previous chapters, nuclear magnetic resonance (NMR) spectroscopy is a powerful tool for the investigation of liquid-crystalline (LC) systems. This is due to the fact that the anisotropic nature of nuclear magnetic dipolar and quadrupolar interactions gives rise to spectral contributions that depend on the molecular orientation with respect to the external static magnetic field, $\mathbf{B}_0$.

In fact, the dipolar splitting of NMR lines associated with proton pairs and the quadrupolar splitting of deuterium in the presence of an electric field gradient (e.g., in a C–D bond) are directly related to the nematic order parameter and the orientation of the director **n**. These splits result from perturbations of the Zeeman levels associated to the interaction between the nuclear spins and the external static NMR magnetic field $\mathbf{B}_0$.

NMR of Liquid Crystal Dendrimers
Carlos R. Cruz, João L. Figueirinhas, and Pedro J. Sebastião

ISBN 978-981-4745-72-7 (Hardcover), 978-981-4745-73-4 (eBook)
www.panstanford.com

Both the dipolar interaction between a pair of spins $I = 1/2$ (typically a proton pair) and the quadrupolar interaction between the spin $I = 1$ (typically a deuteron) and the electric field gradient (EFG) tensor may be described by second-rank tensors as described in Chapter 4.

In the uniaxial nematic phase, the orientational order can be described by a single order parameter, defined by Eq. 2.2. In that case, the physical properties of the system described by second-rank tensors (e.g., the magnetic susceptibility, the electric permittivity, or the refractive index) can be completely defined by two elements, one corresponding to the direction of a principal axis (z), and another one associated to any direction perpendicular to z. If $\mathbf{Q}$ is such a physical parameter it can be expressed in its principal frame (where one of the principal axis coincides with the z direction) by

$$\mathbf{Q} = \begin{bmatrix} Q_\perp & 0 & 0 \\ 0 & Q_\perp & 0 \\ 0 & 0 & Q_\parallel \end{bmatrix} \tag{8.1}$$

where $Q_\parallel$ is the component of $\mathbf{Q}$ in the principal direction (z) and $Q_\perp$ is the value of the component in any direction perpendicular to z. In a uniaxial phase all the directions perpendicular to z are degenerate. The system is symmetric for rotations around the z axis, which in the case of the uniaxial phase, corresponds to the director $\mathbf{n}$. See Fig. 8.1.

In biaxial phases, however, the physical properties depend on the orientation in the plane perpendicular to the principal axis of the phase associated with the nematic director $\mathbf{n}$. The symmetry of rotations around the principal axis is broken and that condition can be described by the emergence of different values for the components of $\mathbf{Q}$ in the x and y directions. As schematically represented in Fig. 8.1, such a nematic phase needs a secondary director ($\mathbf{m}$), perpendicular to the principal director ($\mathbf{n}$) to define the corresponding orientational order. The third director ($\mathbf{l}$) is automatically defined by the direction mutually perpendicular to $\mathbf{n}$ and $\mathbf{m}$. The physical parameter $\mathbf{Q}$ can be represented in its principal frame (with the components defined to fulfill the condition $|Q_{zz}| > |Q_{yy}| \geq |Q_{xx}|$) by

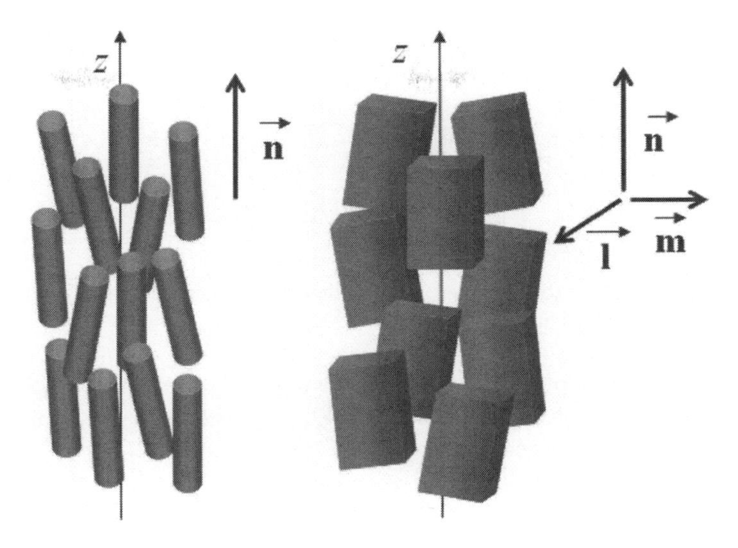

Figure 8.1 Schematic representation of uniaxial (left) and biaxial (right) nematic phases.

$$\mathbf{Q} = \begin{bmatrix} Q_{xx} & 0 & 0 \\ 0 & Q_{yy} & 0 \\ 0 & 0 & Q_{zz} \end{bmatrix} \tag{8.2}$$

with $Q_{xx} \neq Q_{yy}$ contrary to the uniaxial case where $Q_{xx} = Q_{yy} = Q_{\perp}$.

As referred in Chapter 2 the search for the biaxial nematic phase, theoretically predicted in 1970 by Freiser (Freiser, 1970), is still a very topical issue in the field of liquid crystals. Especially in the case of thermotropics, the search for the biaxial nematic phase remains an open problem. After more than thirty years of controversial results (Luckhurst, 2001), by the middle of the 2000 decade, several thermotropic LC systems, namely side-chain polymers (Severing and Saalwachter, 2004), bent-core mesogens (Acharya et al., 2004; Madsen et al., 2004), bent-core dimers (Channabasaveshwar et al., 2004), and organosiloxane tetrapodes (Figueirinhas et al., 2005; Merkel et al., 2004), were claimed to exhibit such a phase.

In principle, any physical parameter described by a second-rank tensor is a good candidate to provide evidence of an eventual biaxial nematic phase. Considering Eqs. 8.1 and 8.2, the measurement of the Q_{xx} and Q_{yy} in the principal frame of the tensorial quantity $\mathbf{Q}$

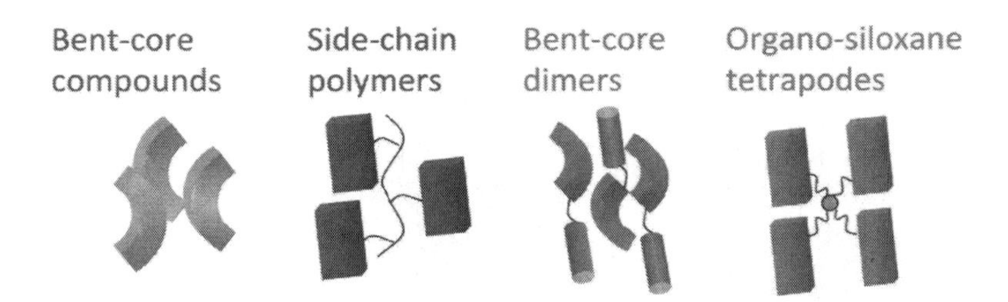

Figure 8.2 Schematic representation of thermotropic molecules exhibiting phases reported as biaxial nematics in 2004–2005.

(with the z axis coincident with the principal direction), is enough to identify the phase as biaxial if $Q_{xx} \neq Q_{yy}$. An *asymmetry* parameter can be defined by the equation

$$\eta = \frac{Q_{yy} - Q_{xx}}{Q_{zz}} \tag{8.3}$$

the phase is recognized as biaxial if $\eta \neq 0$.

Amongst different possible methods, optical techniques, such as polarizing optical microscopy texture defects observations and conoscopy, have been used to identify such a type of phase in LC materials (Luckhurst, 2001; Merkel et al., 2004). However, limitations related to possible biaxiality induced by surface interactions on the LC samples have been frequently questioned in the literature. Contrary, by dealing in general with bulk samples, NMR has been systematically considered as a reference technique to the identification of biaxial nematic behavior in liquid crystals (Galerne, 1988; Luckhurst, 2001; Madsen et al., 2004). Deuterium NMR is particularly useful if partially deuterated samples are used, since each deuterium nucleus in the presence of an electric field gradient (e.g., in a C–D bond) contributes with a single pair of lines to the NMR spectra. Actually, a set of equivalent deuterium nucleus, with the same *average* angle between the C–D bond and the static external magnetic field $\mathbf{B}_0$ gives a single pair of quadrupolar lines. That occurrence is typically identified by the additional amplitude of such a pair of NMR lines. Optionally, if a deuterated sample is not available, mixtures with deuterated materials can be used, assuming that the deuterated molecule used as a *probe* follows the common phase molecular ordering of the *host* material. In the case of deuterium NMR, the physical property **Q** (see Eqs. 8.1, 8.2, and 8.3) used to probe the molecular ordering properties is the electric field gradient (EFG) tensor (**V**) with components V_{ij}.

8.1.2 Organosiloxane Tetrapodes

As referred above, biaxial ordering has been identified in some cases of LC dendrimers, more precisely organosiloxane tetrapodes (generation 0 dendrimers) with laterally attached mesogens. In this chapter, we will address the study of biaxial nematic order in LC dendrimers by means of NMR spectroscopy with particular emphasis on the referred materials (Cardoso et al., 2008; Cruz et al., 2008; Figueirinhas et al., 2005, 2009).

The chemical structure of the organosiloxane tetrapode Tas (a) and the corresponding monomers (b) and (c), used in the referred studies, are represented in Fig. 8.3, together with the respective phases' sequences. In the NMR spectroscopy measurements reported in those references, a mixture (15% weight) with the nematic deuterated probe 7CBαd2 was used. The chemical structure of the probe is presented in Fig. 8.3d. As pointed out before, the experimental procedure is based on the assumption that the molecules of the deuterated probe (in this case compound 7CBαd2) by being mixed with the tetrapodes (or the monomers) assume the nematic ordering of the host material. This technique is currently used in deuterium NMR of liquid crystal phases (Martin-Rapun et al., 2004). The use of a probe that exhibits a nematic phase in its pure form, in the investigations of organosiloxane tetrapodes, was chosen in order to improve the mixing conditions with the host material. In fact, nematic mixtures are generally used in liquid crystals research in order to improve the thermal stability of the materials and their physical properties relevant to the production of technological devices (de Gennes and Prost, 1993).

As shown in the Fig. 8.3, the deuterated probe has two equivalent deuterium nucleus at the α position (first carbon atom) of the aliphatic chain. The quadrupolar NMR spectra described in the next sections, correspond to a single pair of lines associated with these two ^{2}H nucleus. In one of the works referred (Cardoso et al., 2008), the NMR spectra of the tetrapodes are compared with those

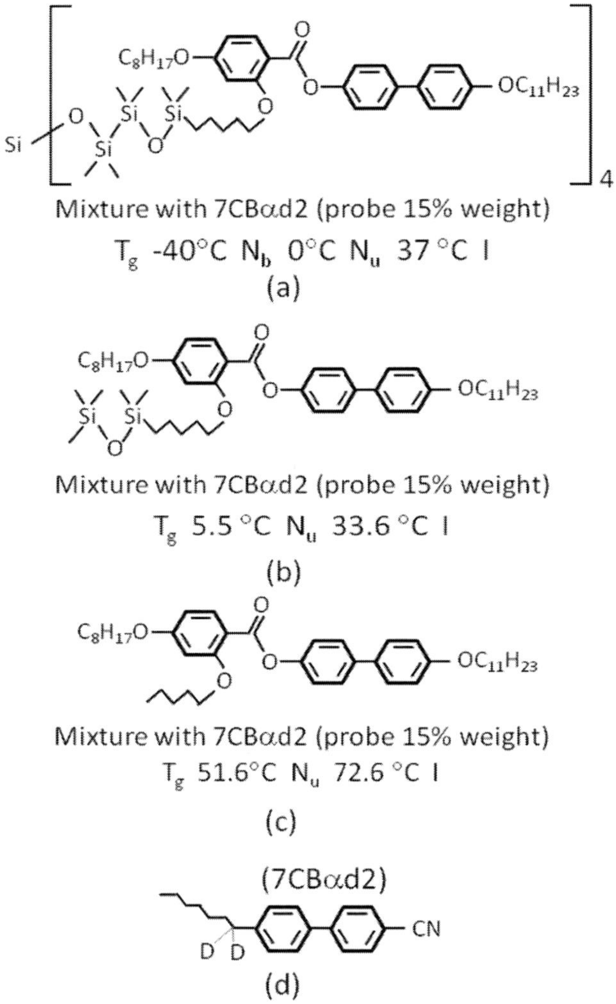
C8H17O
OC11H23
Si
4
Mixture with 7CBαd2 (probe 15% weight)
Tg -40°C Nb 0°C Nu 37 °C I
(a)
C8H17O
OC11H23
Mixture with 7CBαd2 (probe 15% weight)
Tg 5.5 °C Nu 33.6 °C I
(b)
C8H17O
OC11H23
Mixture with 7CBαd2 (probe 15% weight)
Tg 51.6°C Nu 72.6 °C I
(c)
(7CBαd2)
CN
D D
(d)

Figure 8.3 Chemical structure and phases' sequences of organosiloxane tetrapodes exhibiting the biaxial nematic phase (a), respective monomers (b, c), and the nematic LC deuterated probe 7CBαd2 (d).

of the corresponding monomers. It is interesting to notice that the difference between the chemical structure of tetrapode and the monomers is the linkage to the dendritic core, corresponding to the central silicon atom. Nevertheless, the NMR spectroscopy results of the monomers nematic phase are completely different from those of the tetrapodes as shown in the next sections. This result is a very clear indication of the influence of the dendritic molecular structure on the molecular ordering. In Fig. 8.4 a

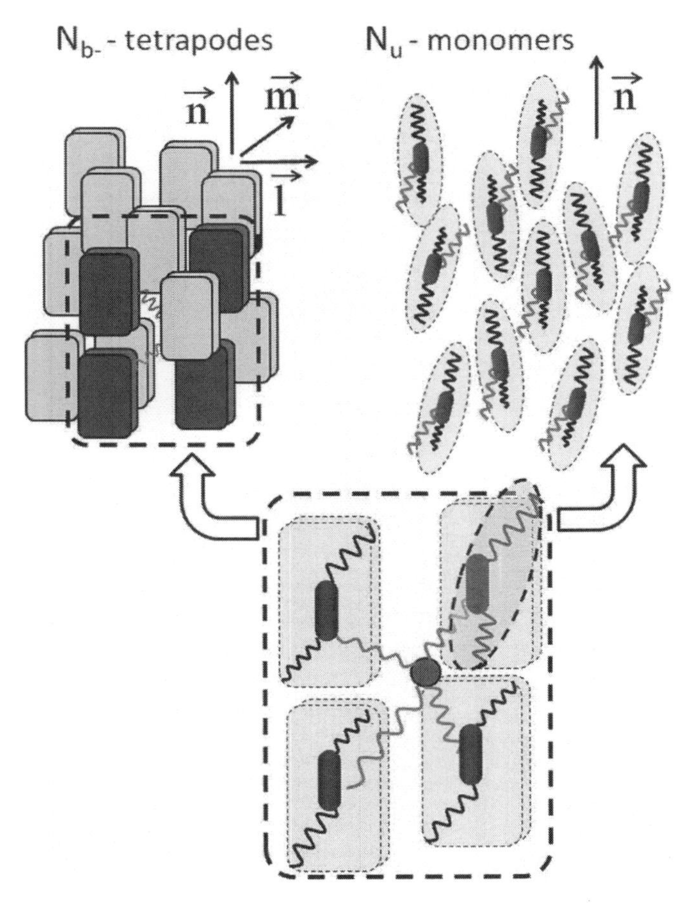

Figure 8.4 Schematic view of the N_b phase of the organosiloxane tetrapodes and N_u phase of the respective monomers.

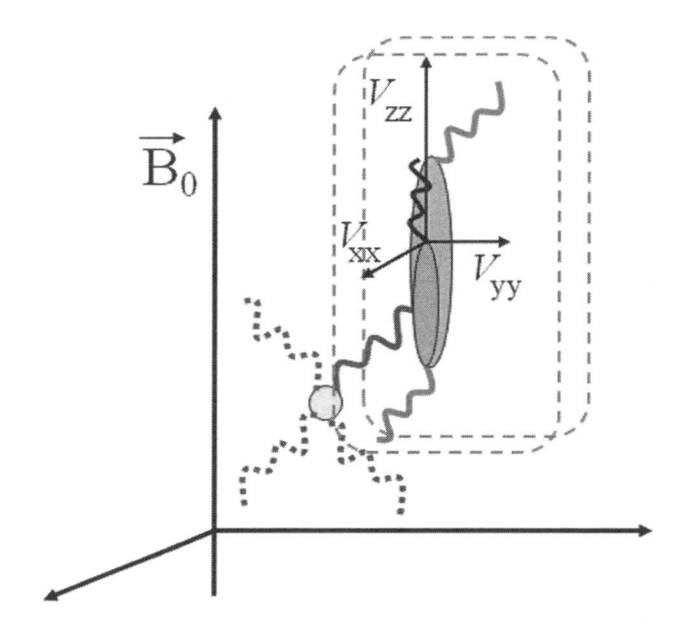

Figure 8.5 Schematic view of the alignment of the deuterated probe with respect to the tetrapode arm or monomer in a domain aligned with the external static magnetic field $\mathbf{B}_0$.

schematic view of the nematic phases of those systems is shown. Similarly, the molecular dynamics of the tetrapodes also reflects the influence of the dendritic structure when compared with that of the corresponding monomers, as shown in Chapter 9. Further details on the Tas dendrimer and other similar systems, regarding the phases' organization and molecular dynamics are also presented in 9. Here we will focus on the molecular order in the nematic phases of Tas and related monomers and the investigation of biaxiality.

In Fig. 8.5 a schematic view of the orientation of the deuterated probe with respect to the tetrapode arm or corresponding monomer is presented. In a homogeneous nematic mixture, a common director represents the preferred orientation of the molecules of the components. In the present case, the main molecular axis of the probe is expected to align with the principal director of the tetrapodes nematic phase. The figure corresponds to the case in which the nematic director is aligned with the NMR external static magnetic field $\mathbf{B}_0$.

8.1.3 Experimental Procedures

As presented in previous chapters, the quadrupolar splitting corresponding to a C–D bond in the presence of an external static magnetic field $\mathbf{B}_0$ is given by (see Eqs. 5.5 and 5.34):

$$\delta\omega = \frac{3\pi}{2}\bar{\nu}_Q\left[(3\cos^2\theta - 1) + \eta\sin^2\theta\cos(2\varphi)\right] \tag{8.4}$$

where $\bar{\nu}_Q$ is the average quadrupolar coupling constant in the principal frame of the average *EFG* tensor associated with the C–D bond, $\bar{\nu}_Q = \frac{eQ}{h}\bar{V}_{\zeta\zeta}$, θ and φ are Euler angles defining the orientation of the NMR static magnetic field $\mathbf{B}_0$ in the principal frame of the average electric field gradient tensor $\bar{V}$. The asymmetry parameter η is given by

$$\eta = \frac{\bar{V}^{\xi\xi} - \bar{V}^{\gamma\gamma}}{\bar{V}^{\zeta\zeta}} \tag{8.5}$$

where $\bar{V}^{\xi\xi}$, $\bar{V}^{\gamma\gamma}$, and $\bar{V}^{\zeta\zeta}$ are the averaged components of the *EFG* tensor in the principal frame.

In a nematic domain aligned with the NMR static field direction z, the ζ axis of the principal frame of the average *EFG* tensor (coincident with the principal director $\mathbf{n}$) is parallel to z. In that condition (corresponding to $\theta = 0$), taking into account Eq. 8.4 and Fig. 8.6, it's easy to verify that the quadrupolar splitting is given by $\delta\omega = 3\pi\,\bar{\nu}_Q$, and will be independent of the asymmetry parameter η and of the orientation of any hypothetical secondary director that would be defined (by the angle φ) in the case of a biaxial phase. Therefore, the determination of a deuterium spectrum of an aligned nematic phase gives no information of a possible biaxial ordering of such a phase. In order to obtain experimental information on the asymmetry parameter η, which corresponds to the signature of biaxiality it is necessary to deal with spectra of nonaligned samples ($\theta \neq 0$). Two possible methods presented in the literature and used in NMR spectroscopy studies of biaxiality ordering, correspond to successive 90° rotation and continuous rotation of the samples, respectively (Luckhurst, 2001).

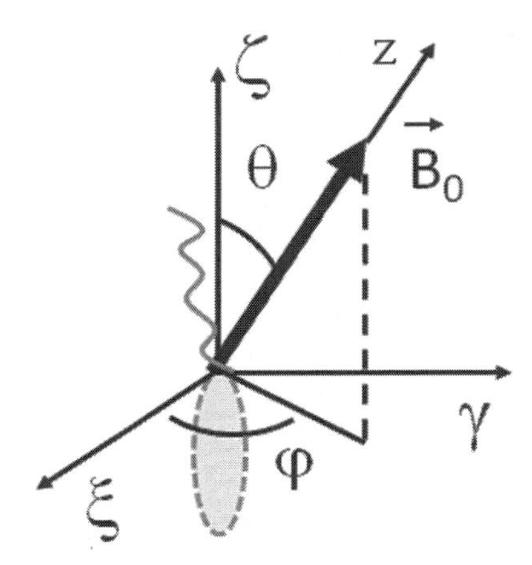

Figure 8.6 Schematic view of the principal frame of the average electric field gradient (*EFG*) tensor (ξ, γ, ζ) and the Euler angles, θ and φ, which define the orientation of the external static magnetic field $\mathbf{B}_0$ with respect to the *EFG* principal frame.

8.1.3.1 90° rotation

Taking an initially aligned ($\theta = 0$) uniaxial ($\eta = 0$) monodomain nematic sample, characterized by a single deuterium line pair, it is immediate to verify, considering Eq. 8.4, that the corresponding quadrupolar splitting is reduced by a factor 1/2 when the sample is rotated by 90° around an axis perpendicular to the NMR static magnetic field $\mathbf{B}_0$ ($\theta = 90°$). If the spectrum is characterized by more than one pair of lines, the same happens to each pair, so the whole spectrum is "compressed" by a 1/2 factor.

Something quite different occurs if we are dealing with a biaxial phase. Taking into account that $\eta \neq 0$, the factor multiplying the line splitting of the 90° rotated sample (1/2 in the case of a uniaxial domain) will include an additional term $\eta \cos(2\varphi)$. This way, if the angle φ, characterizing the orientation of the nematic secondary director, is known, the value of the asymmetry parameter η can be directly measured from such an experiment.

In real laboratory conditions, however, different types of limitations to that ideal situation arise, making such an experiment not so straightforward. The first, and possibly the most obvious of all, lays on the fact that, in a hypothetical biaxial nematic sample

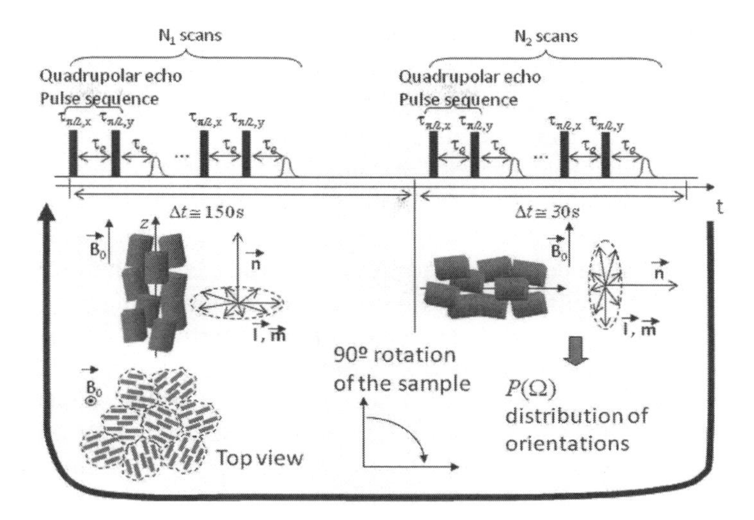

Figure 8.7 Schematic view of the deuterium NMR spectroscopy 90° rotation experiment.

with the principal director **n** aligned with the NMR magnetic field $\mathbf{B}_0$, there is no physical reason for the secondary director to be aligned. Therefore, in principle, such a sample will be formed by a polydomain of nematic clusters with the principal director aligned with $\mathbf{B}_0$ and the secondary directors uniformly distributed in the plane perpendicular to $\mathbf{B}_0$, as depicted in Fig. 8.7. This initial situation corresponds to a uniform distribution of the angle φ. Therefore, in general, the spectrum of a real biaxial 90° rotated sample will be more complex than the simple one resulting from the rotation of the ideal monodomain sample.

On the other hand, after the 90° rotation, the principal director tends to revert to the initial direction, due to the effect of the magnetic torque exerted by $\mathbf{B}_0$ on the molecules. This way, the 90° rotation technique requires the viscosity of the nematic sample to be high enough so that the duration of the NMR spectra acquisition is much shorter than the reorientation time of the principal nematic director. This draw back excludes the use of this technique on low molecular weight liquid crystals that have reorientation times of the order of the millisecond. Even in the highly viscous nematic polymers, with long reorientation times (of the order of seconds or longer), special attention must be paid to this problem. Generally, the achievement of reasonable signal-to-noise ratios of the NMR signal requires the accumulation of a significant number of scans. That process takes a considerable time that would make the technique eventually unusable, even for the nematic polymers, unless some additional procedure is applied.

In order to overcome this limitation, a sequence of repeated 90° rotations separated by short acquisition periods, in which the NMR scans are accumulated, is applied. Of course the acquisition periods must be much shorter than the principal director reorientation time of the material under study (Severing and Saalwachter, 2004). This method is schematically described in Fig. 8.7. The sample is initially aligned with the principal director $\mathbf{n}$ parallel to $\mathbf{B}_0$ and a quadrupolar echo pulse sequence (described in a previous chapter) is applied and N_1 scans are accumulated in that condition, the sample then is rotated by 90° around an axis perpendicular to $\mathbf{B}_0$ and a number N_2 of scans is accumulated, using again the quadrupolar echo sequence. The time Δt must be short enough in order to assure that the principal director keeps the $\theta = 90°$ orientation during the acquisition process. Next, the sample reverts to the original aligned position $\theta = 0$ and the process is repeated as many times as necessary to obtain a sum of accumulated signals with a good signal to noise ratio. The time values presented in Fig. 8.7 correspond to the experiments with organosiloxane tetrapodes described in references (Figueirinhas et al., 2005) and (Cruz et al., 2008).

This method has several advantages. On the one hand, by going back repeatedly to the initial oriented position of the main director ($\theta = 0$) the sample preserves its orientational order. This condition is straightforwardly controllable by accumulating a number of scans at the ($\theta = 0$) position and comparing the respective spectrum with that of a static oriented sample with the principal director parallel to $\mathbf{B}_0$. The 90° orientation can also be directly verified by measuring the quadrupolar splitting corresponding to the 90° rotated sample.

On the other hand, when the principal director is repeatedly rotated by 90°, one of the secondary directors will be potentially aligned parallel to $\mathbf{B}_0$ (and the other perpendicularly) due to the associated magnetic torque. This process will eventually occur, producing a total or partial alignment of the secondary directors and its effect on the NMR spectra is determined by the analysis of the 90° rotated spectra.

That technique was applied to the detection of the biaxial nematic ordering in thermotropic side-chain polymers (Severing and Saalwachter, 2004) and generation 0 dendrimers (organosiloxane tetrapodes) (Cruz et al., 2008; Figueirinhas et al., 2005).

8.1.3.2 Continuous rotation

Another way to detect the possible biaxial ordering of a nematic liquid crystal is the accumulation of a sequence of scans while the sample is continuously rotated around an axis perpendicular to $\mathbf{B}_0$. This process is schematically depicted in Fig. 8.8 and its application is described in the literature for instance in NMR spectroscopy measurements on the nematic phases (Madsen et al., 2004) of bent-core compounds. Ideally, the continuous rotation of the sample gives rise to a 2D nematic powder. Under these circumstances, the main director becomes uniformly distributed in the rotation plane and eventually one of the secondary directors will tend to align (at least partially) with the rotation axis perpendicular to $\mathbf{B}_0$.

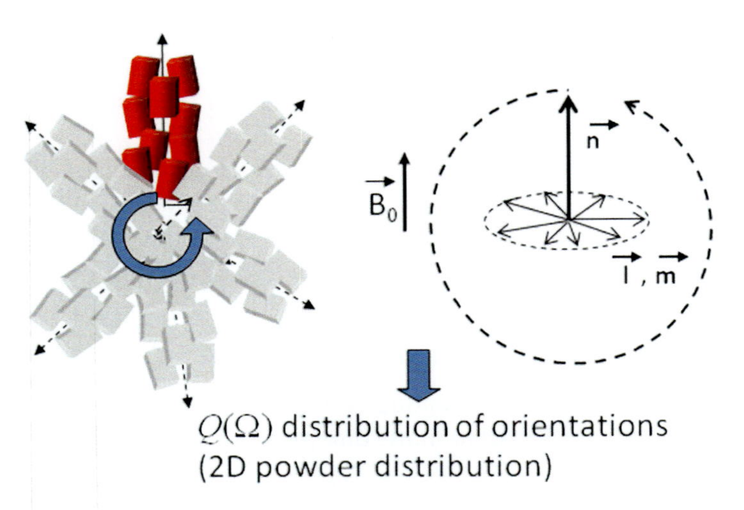

Figure 8.8 Schematic view of the continuous rotation of a nematic sample in a deuterium NMR spectroscopy experiment.

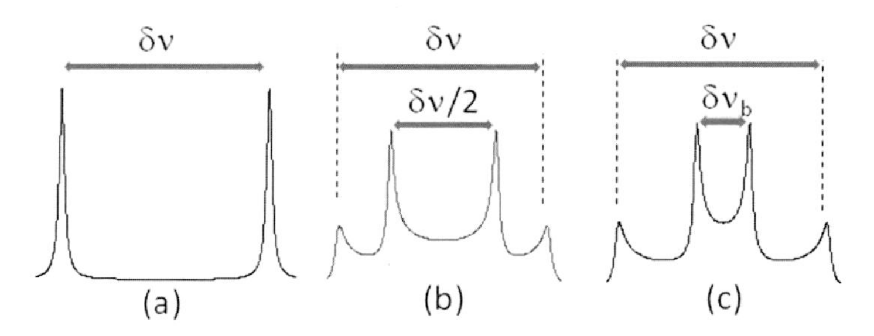

Figure 8.9 Deuterium spectra of a nematic sample: (a) Sample aligned with $\mathbf{B}_0$, (b) 2D powder sample of uniaxial nematic domains, and (c) 2D powder sample of biaxial nematic domains ($\delta\nu_b \neq \delta\nu/2$).

The spectrum of such a "polycrystalline" sample is the result of the sum of a distribution $Q(\Omega) = Q(\theta, \varphi)$ of line pairs with individual splits determined by the Euler angles θ, and φ corresponding to the relative orientation of $\mathbf{B}_0$ in the principal frame of the average *EFG* tensor. A more detailed description of the spectra simulation is presented in the next subsection.

The spectrum of an aligned sample ($\theta = 0$) with a single deuterium pair of lines is shown in Fig. 8.9a.

The spectrum of a 2D uniaxial nematic powder has a distinct shape, shown in Fig. 8.9b, presenting two pairs of "peaks" with the "external" pair (corresponding to $\theta = 0$) characterized by a splitting which is the double of that of the internal one (corresponding to $\theta = 90^\circ$).

In the case of a biaxial nematic phase (Fig. 8.9c), the shape of the 2D powder spectrum is different and immediately identifiable since the ratio between the splits of external and internal peak pairs is different from two.

8.1.4 Experimental Results and Spectra Simulations

8.1.4.1 90° rotation technique and spectra of a static distribution of nematic domains

The 90° rotation technique was used in a sample of the organosiloxane tetrapode Tas, mixed with the nematic deuterated probe 7CBαd2. The chemical structure of Tas and of the probe and the phases' sequence of the mixture (15% weight) is shown in Fig. 8.3. The isotropic-nematic phase transition is detected both by polarizing optical microscopy and differential scanning calorimetry (DSC). The Nu-Nb phase transition in the mixture is observed by deuterium NMR spectroscopy, as discussed below (Cruz et al., 2008; Figueirinhas et al., 2005). The phases' sequence of the pure compound Tas is shown in Fig. 9.4 (Filip et al., 2010).

The deuterium spectra, obtained experimentally by means of the 90° rotation technique, on the Tas compound, using the deuterated probe are presented in Fig. 8.10a,b. The respective theoretical simulations, that allow for the determination of the most relevant physical parameters in the problem ($\bar{\nu}_Q$ and η), are also shown in the figure (red line).

In the case of the 90° rotation technique the simulated spectra can be described by the function

$$G(\omega) = \int_{\Omega} \left[L\left(\omega - \frac{\delta\omega}{2}\right) + L\left(\omega + \frac{\delta\omega}{2}\right) \right] P(\Omega)\, d\Omega \qquad (8.6)$$

with $\delta\omega$ given by Eq. 8.4 and $P(\Omega)$ ($\Omega = (\theta, \varphi)$) accounts for the distribution of secondary directors discussed above, and $L(\omega)$ is an appropriate function that describes the NMR line shape (typically a Lorenzian function):

$$L(x) = \frac{1}{1 + \left(\frac{x}{\lambda}\right)^2} \qquad (8.7)$$

where λ is the half-width at half-maximum.

8.1.4.2 Continuous-rotation technique and the effect of slow motions

Given the results obtained with the continuous-rotation technique (Fig. 8.10c) it became obvious, from the collected spectra, that a more complex theoretical approach would have to be considered in order to obtain reasonable fittings to the experimental data. Contrary to what is found in the data presented in Fig. 8.10c, it is known that typical spectra of 2D powder samples, also in biaxial phases, present a pronounced pair of peaks in the central region that doesn't appear in this case for the lowest temperatures (see Figs. 8.9 and 8.10c). A detailed discussion on this question is presented in reference (Figueirinhas et al., 2009).

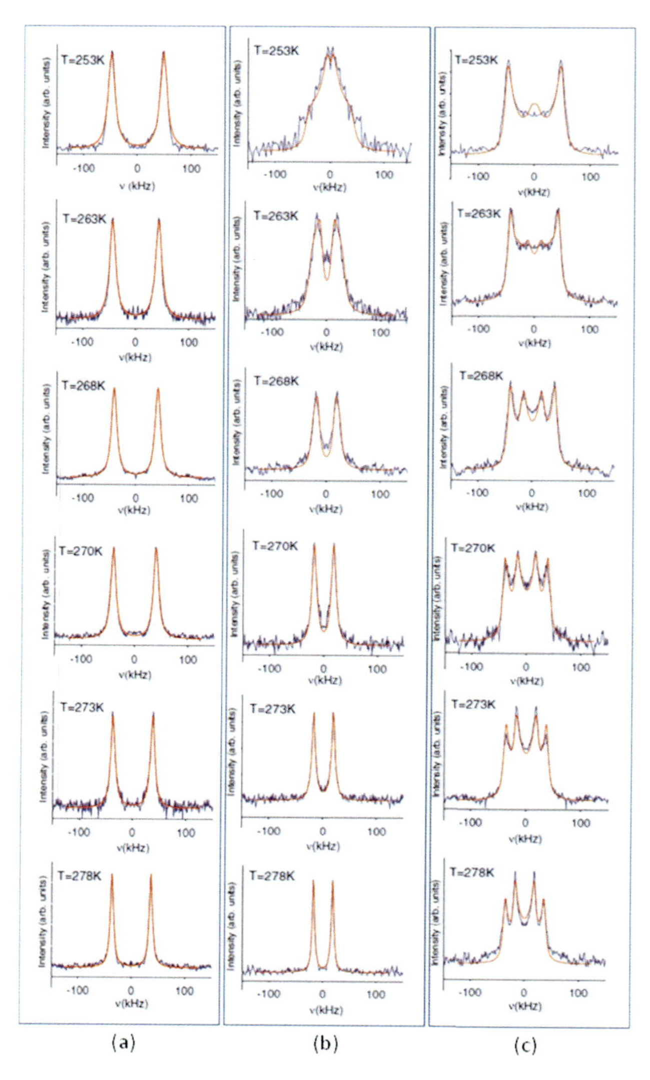

Figure 8.10 Deuterium spectra of Tas mixture with the 7CBαd2 probe (15% weight). Experimental data, blue line; simulated spectra, red line. (a) Sample aligned with $\mathbf{B}_0$, (b) sample rotated 90° with respect to $\mathbf{B}_0$, and (c) continuous rotation. Reprinted (figure) with permission from [Cruz, C., Figueirinhas, J. L., Filip, D., Feio, G., Ribeiro, A. C., Frere, Y., Meyer, T. and Mehl, G. H. (2008). Biaxial nematic order and phase behavior studies in an organosiloxane tetrapode using complementary deuterium NMR experiments, *Physical Review E* **78**, 5, p. 051702]. Copyright (2008) by the American Physical Society.

An explanation for this experimental fact may be found in the effect of molecular *slow motions*. If the collective movements of the molecules are slow compared with the NMR observation time (typically of the order of the microsecond and given approximately by the inverse of the spectra bandwidth) the NMR signal obtained during a single acquisition is affected by that molecular movement giving rise to the broadening of the line. The reason why this effect doesn't influence significantly the data in the case of the 90° rotation technique is most probably due to the fact that the line splitting function (Eq. 8.4) is stationary both for $\theta = 0$ and $\theta = 90^\circ$. Therefore, the quadrupolar splitting $\delta\omega$ doesn't suffer considerable changes due the fluctuations of the principal director when it is aligned parallel or perpendicular to $\mathbf{B}_0$. For other director orientations, however, this effect may be considerable as shown by simulations reported in (Cruz et al., 2008) and (Figueirinhas et al., 2009).

The simulated spectra $G(\Omega)$ is given by the Fourier transform of the NMR *FID* signal:

$$G(\omega) = \mathrm{TF}\,[g(t)] \tag{8.8}$$

with

$$g(t, \Omega) = \int_{\Omega} f(t, \Omega)\, Q(\Omega)\, d\Omega \tag{8.9}$$

where $g(t)$ is the *FID* signal and, in the case of continuous rotations, the distribution of orientations $Q(\Omega)$ accounts for the rotation of the sample (which is equivalent to the rotation of $\mathbf{B}_0$ in the principal frame of the *EFG* tensor).

The function $f(t, \Omega)$ associated with a nematic monodomain can be obtained considering the result of the application of the quadrupolar echo pulse function described before:

$$f(t, \Omega) \propto \mathrm{Re}\left\langle \exp\left\{ i\int_{\tau}^{t+2\tau} \omega_0[\Omega, t', r]\, dt' - i\int_{0}^{\tau} \omega_0[\Omega, t', r]\, dt' \right\}\right\rangle \tag{8.10}$$

where Re stands for the real part of, the brackets $\langle \ldots \rangle$ indicate an ensemble average, τ is the time between NMR pulses in the quadrupolar echo sequence, and ω_0 is the time dependent absorption frequency of the nuclear spins, associated to the quadrupolar interaction, at time t' at $\mathbf{r}$ (position corresponding to the nematic domain) given by

$$\omega_0(\alpha_1, \alpha_2, t', \mathbf{r}) = \frac{3\pi}{\sqrt{6}} \frac{eQ}{h} \sum_{m=-2,n=-2}^{2,2} \mathbf{D}^2_{m,0}(\alpha_2, \pi/2, 0) \mathbf{D}^2_{n,m}(\alpha_1, \pi/2, \pi/2) V_n(t', \mathbf{r}) \qquad (8.11)$$

where e is the electron charge, Q is the quadrupolar moment of the deuterium nucleus, h is the Planck constant, $\mathbf{D}^2_{ij}$ are the components of the second-rank Wigner rotation matrices (Abragam, 1961), V_n are the irreducible components of the time-dependent *EFG* tensor in the nematic domain at position $\mathbf{r}$, and the angles α_1 and α_2 define the orientation of the NMR static field $\mathbf{B}_0$ in the principal frame of the average electric field gradient tensor in the nematic domain (see Fig. 8.11).

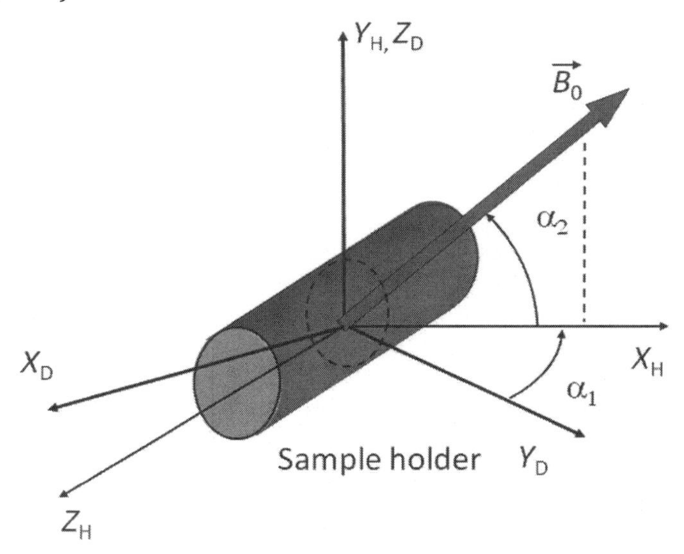

Figure 8.11 Schematic representation of the sample holder and definition of the angles α_1 and α_2 that describe the orientation of the static magnetic field $\mathbf{B}_0$ in the principal frame of the *EFG* tensor in each nematic domain (frame D). The rotation of the sample holder in the laboratory frame is associated with angle α_2 and the position of the principal frame of the *EFG* tensor in each nematic domain with respect to the sample holder frame is defined by α_1.

For the simulation of the NMR spectra (see Eqs. 8.8 and 8.9) it is necessary to evaluate the ensemble average in Eq. 8.10. To that purpose, ω_0 is considered as a Gaussian random variable due to the fluctuations that affect $V_n(t', \mathbf{r})$.

This calculation involves the determination of the average $\langle V_n(t', \mathbf{r}')\rangle$ and correlation functions proportional to $\langle V_n(0, \mathbf{r})$ $V_n(t', \mathbf{r}')\rangle$ for different nematic monodomains defined by Eq. 8.11. To this purpose the irreducible components of the *EFG* tensor $V_n(t', \mathbf{r}')$ will be considered to be proportional to those of the tensor order parameter $Q_n(t', \mathbf{r}')$

$$V_n(t', \mathbf{r}') = c\, Q_n(t', \mathbf{r}') \tag{8.12}$$

and, in order to account for slow motions, the components of the order parameter tensor are decomposed into an average part Q_n^0 and an additional term associated with fluctuations due to collective modes $\Delta Q_n((t', \mathbf{r}')$

$$Q_n(t', \mathbf{r}') = Q_n^0 + \Delta Q_n(t', \mathbf{r}') \tag{8.13}$$

with

$$\begin{aligned} Q_{\pm 2}^0 &= \frac{\eta S}{2} \\ Q_{\pm 1}^0 &= 0 \\ Q_0^0 &= \frac{\sqrt{6}}{2} S \end{aligned} \tag{8.14}$$

where η and S (the asymmetry parameter and the nematic order parameter, respectively) are the fitting parameters of the spectra simulation.

The constant c in Eq. 8.12 can be determined considering that, by definition

$$V_n^0 \equiv \left\langle V_n(t', \mathbf{r}')\right\rangle = c\, Q_0^n \tag{8.15}$$

the quadrupolar coupling constant is given by

$$\nu_Q = \frac{eQ}{h} V_{zz}^0 \tag{8.16}$$

and using the component with $n = 0$, and V_{zz}^0 from Eq. 8.16

$$c = \frac{V_0^0}{Q_0^0} = \frac{V_{zz}^0}{S} = \frac{h}{eQ}\frac{\nu_Q}{S} = \frac{h}{eQ}\frac{\nu_{QS}}{1} \tag{8.17}$$

The parameter $\nu_{QS} = \nu_Q/S$ can be estimated from the value of the quadrupolar coupling constant for the solid state at the α position of the aliphatic chain and the angle between the C–D bond and the para-axis of the benzene ring reported in the literature. This values and all further details on the calculations, including the comprehensive mathematical approach used in the description of the collective modes fluctuations, are presented in reference (Cruz et al., 2008).

8.1.5 Discussion of the Spectra Simulation Results

The results of the spectra simulations are presented in Fig. 8.10c and the corresponding physical parameters obtained by the combined analysis of the continuous-rotation and 90° rotation techniques, ν_Q, S and η are shown in Fig. 8.12.

The results of the asymmetry parameter, shown in Fig. 8.12b clearly indicate that the system exhibits biaxial ordering for temperatures below 0° (273K). The difference between the N_u-N_b transition temperature found by NMR (Cruz et al., 2008; Figueirinhas et al., 2005) and that found by Vij and co-workers in 2004 (Merkel et al., 2004) by optical and infrared spectroscopy experiments is most probably due to the fact that the NMR experiments are done with a mixture (with the deuterated probe 7CBαd2). The shift of the phase transition temperatures in LC mixtures in comparison with pure compounds is normally observed. In the case discussed herein, the transition temperatures, both in the mixture and in the pure compounds (Merkel et al., 2004) are shown in Figs. 8.3 and 9.4, respectively. The results regarding the biaxial ordering of the system are immediately evident from the observation of the 90° rotation experimental results as shown in Fig. 8.10a,b. The results for upper temperatures are typical of uniaxial nematic phases, showing a contraction of 1/2 for the quadrupolar splitting when the sample is rotated by 90° (as explained before and easily concluded from Eq. 8.4). However, that is strikingly not the case for lower temperatures. From this experiment alone it is possible obtain the values of η, S and ν_Q. The rotating sample experiment provides a reliable confirmation of the 90° rotation results, since the spectra of 2D powder samples (see Fig. 8.10c) are correctly fitted with the

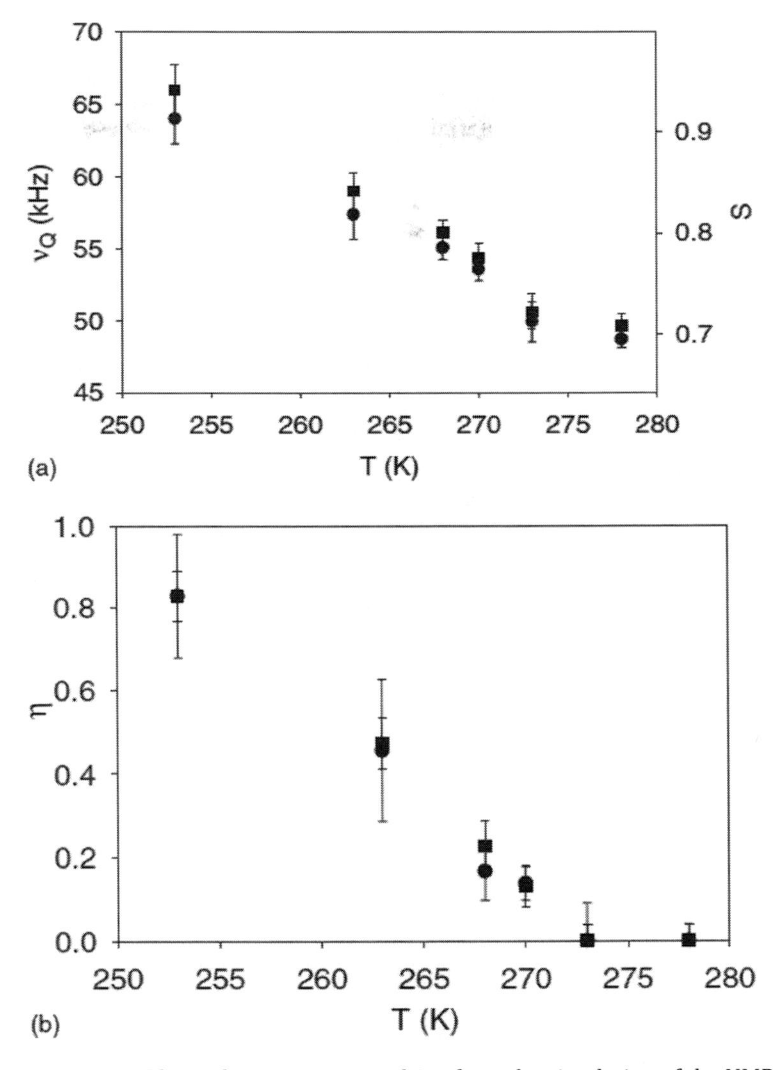

Figure 8.12 Physical parameters resulting from the simulation of the NMR deuterium spectra obtained as a function of temperature, combining the fits of theoretical expressions to the experimental data collected both from the 90° rotation and continuous-rotation techniques: (a) ν_Q and S; (b) η. Reprinted (figure) with permission from [Cruz, C., Figueirinhas, J. L., Filip, D., Feio, G., Ribeiro, A. C., Frere, Y., Meyer, T. and Mehl, G. H. (2008). Biaxial nematic order and phase behavior studies in an organosiloxane tetrapode using complementary deuterium NMR experiments, *Physical Review E* **78**, 5, p. 051702]. Copyright (2008) by the American Physical Society.

same physical parameters obtained from a completely independent experiment (90° rotation).

The simulation of the spectra involves the consideration of a partial orientation distribution of the secondary directors **l** and **m** characterized by the angle φ (see Eq. 8.4 and Fig. 8.6), both for the 90° rotation and continuous-rotation experiments.

The distribution $P(\Omega)$ for a sample aligned with the NMR static magnetic field $\mathbf{B}_0$ is given by:

$$P_0(\Omega) = \frac{1}{\pi}\delta(\cos\theta - 1) \tag{8.18}$$

This expression corresponds to a uniform distribution in the angle φ (associated with the orientation of the secondary directors) when the principal director is aligned with $\mathbf{B}_0$ (see Fig. 8.7).

When the principal director in the sample, initially aligned, is rotated by 90° with respect to the static NMR field $\mathbf{B}_0$, the directors distribution becomes

$$P(\Omega) = \delta(\cos\theta)h(\varphi) \tag{8.19}$$

where $h(\varphi)$ is given by

$$h(\varphi) = \begin{cases} [1 + C_1\cos(2\varphi) + C_2\cos(4\varphi) + C_3\cos(6\varphi)]\,/C_0, & h(\varphi) \geq 0 \\ 0, & h(\varphi) < 0 \end{cases} \tag{8.20}$$

C_0 is a normalization constant and C_1, C_2 and C_3 are fitting parameters that account for the partially alignment of the secondary directors.

In the ideal situation of a perfectly aligned biaxial nematic phase, the function $h(\varphi)$ would be given by

$$h(\varphi) = 1/2\delta[\cos^2(\varphi) - 1] \tag{8.21}$$

In the case of the studied system, the results of $h(\varphi)$ at different temperatures for the 90° rotation and continuous-rotation experiments are given in Fig. 8.13a and Fig. 8.13b, respectively.

As can be seen in the figure, the partial orientation distribution of secondary directors change with the technique used (90° versus continuous rotation) and with the temperature. This is not unexpected since the experimental conditions are different with respect to the magnetic torque applied to the molecules or molecular segments (dendritic arms) in the different angular conditions. On the

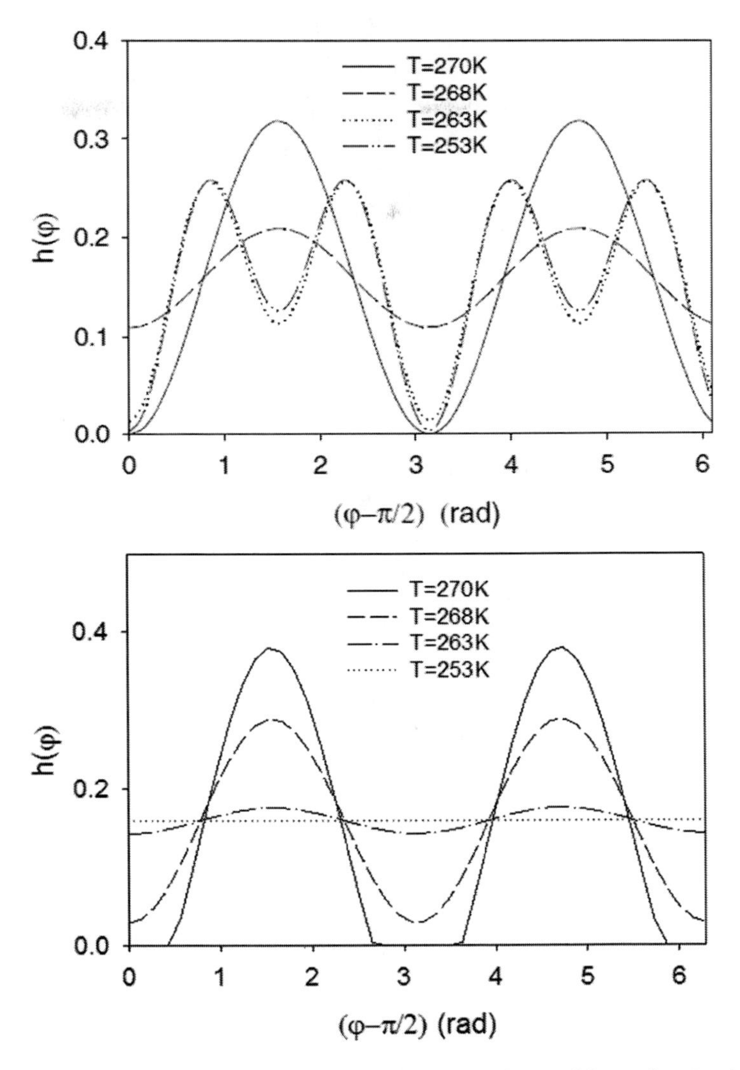

Figure 8.13 Distribution of secondary directors obtained from the simulation of the NMR deuterium spectra as a function of temperature, combining the fits of theoretical expressions to the experimental data collected using (a) the 90° rotation technique and (b) the continuous-rotation technique. Reprinted (figure) with permission from [Cruz, C., Figueirinhas, J. L., Filip, D., Feio, G., Ribeiro, A. C., Frere, Y., Meyer, T. and Mehl, G. H. (2008). Biaxial nematic order and phase behavior studies in an organosiloxane tetrapode using complementary deuterium NMR experiments, *Physical Review E* **78**, 5, p. 051702]. Copyright (2008) by the American Physical Society.

other hand, the degree of alignment of the secondary directors within the biaxial nematic domains varies with the temperature. This results can be explained by the competition between the expected increase of the material's viscosity with decreasing temperatures with the opposite tendency of alignment of secondary directors with the increasing of the asymmetry parameter η for lower temperatures.

8.2 Comparison with Tetrapodes' Monomers and Other Main-Chain LC Dendrimers

The biaxial nematic ordering detected in the organosiloxane tetrapodes by NMR, was also observed by means of several different experimental techniques, namely IR spectroscopy and optical studies (Merkel et al., 2004; Polineni et al., 2013) and dynamic light scattering in a similar compound with a germanium core (Neupane et al., 2006). However, the N_u-N_b was not detected by high resolution differential scanning calorimetry (DSC) (Cordoyiannis et al., 2008) and other techniques involving optical observations (Kim et al., 2014) and capacitance measurements of electro-optical cells (Polineni et al., 2013).

The conclusions to be taken from these discrepancies may be related to the complex molecular structure of these materials and their intrinsic formation of nematic clusters with biaxial arrangement (Filip et al., 2010). According to a theory, developed by Photinos and co-workers, a nematic cluster phase is formed by a set of biaxial nematic clusters. When the principal director, **n** of the different clusters is commonly aligned, the secondary directors remain randomly distributed in the plane perpendicular to **n**, forming a 2D biaxial nematic powder. The secondary directors can be subsequently aligned by a field perpendicular to the main director, forming an aligned biaxial monodomain induced by the external field (Peroukidis et al., 2009; Vanakaras and Photinos, 2008). This type of behavior would explain why, under certain circumstances, the biaxial nematic ordering is not observed, whilst it is so clearly observed by other experimental techniques, namely

NMR, where a particularly intense magnetic field is involved. It is worthwhile to notice that the same type of questions were raised about biaxial nematic ordering observations in bent-core mesogens and explained considering the emergence of macroscopic biaxiality in cybotactic biaxial nematic molecular organization under the effect of external fields (Francescangeli and Samulski, 2010; Samulski, 2010).

The possibility of biaxial nematic ordering was also investigated in generation 3 DAB codendrimers (bearing 32 terminal groups) with laterally attached mesogenic units, synthesized by Serrano and co-workers in the University of Zaragoza, in a study in collaboration with Luckhurst and co-workers at the University of Southampton (Martin-Rapun et al., 2004). Those studies were conducted on the codendrimer $DL_{24}T_8$ bearing 24 laterally attached and 8 terminally attached mesogenic units, respectively. The 90° rotation technique, described before, was applied in that work. For the acquisition of deuterium NMR spectra, the deuterated probe p-xylene-d10 at 2.5% weight, dissolved in the codendrimer, was used. It was observed that the line splittings of the deuterium NMR spectra were reduced to 1/2, when the sample was rotated by 90° with respect to the director **n**. As discussed in Section 8.1.3.1 this result is fully consistent with the uniaxial nematic phase structure. This result is not really unexpected since an eventual biaxial arrangement would imply a coordinated orientation of the transverse axes of mesogenic units. In the present case, the large dendritic core with 32 functional units can be arranged, due to its flexibility, in order to fulfill the condition of common orientation of the longitudinal axes of the terminal units. However, the correlation between the transverse molecular axes of the mesogenic terminal dendritic units (even those laterally attached) would be unexpected since, in the same dendrimer, they are separated by the large flexible dendritic core. It could be argued that, the biaxial nematic phase is observed in the case of polymers with laterally attached mesogenic units (Severing and Saalwachter, 2004). However, in the case of polymers the chain is not at all so bulky as the dendritic core of a G3 dendrimer. Therefore, the laterally attached mesogenic units in the polymer can interact closely to each other in order to align their secondary (transverse) molecular axes when the temperature is lowered and give rise to the biaxial nematic phase. A similar effect is observed in the case of organosiloxane tetrapodes (generation 0 dendrimers) where the dendritic core is minimal and all the mesogenic units in a dendrimer are packed close to each other and interdigitated with the neighboring molecules (Filip et al., 2010).

Finally, in order to investigate the role of the connection between the mesogenic units and the dendritic core in the uprising of the biaxial nematic ordering, a study was conducted using the exact monomers that form the organosiloxane tetrapodes and an identical mixture with the same deuterated probe used in the case of the tetrapodes (Fig. 8.3d). The molecular structure and phases' sequence of the materials are presented in Fig. 8.3 (Cardoso et al., 2008). In the tetrapodes the dendritic core is reduced to the central silicon atom, or the central silicon plus the four siloxane chains, according to the defined convention. Two monomers were used in that investigation: (i) a completely laterally substituted monomer with the exact molecular structure of the dendrimer arm (Fig. 8.3b); (ii) a nonsubstituted monomer corresponding basically to the mesogenic unit with a lateral aliphatic spacer (Fig. 8.3c).

Contrary to the tetrapodes, the low viscosity of the nematic phases formed by the monomers gives rise to a fast reorientation process when the director **n** of an aligned sample is rotated away from the NMR magnetic field direction. This fact excludes the use of the 90° rotation technique. Therefore, the continuous-rotation technique was used for the investigation of eventual biaxial nematic behavior of both materials.

The results and experimental procedure are discussed in detail in (Cardoso et al., 2008), and briefly presented in Fig. 8.14.

The spectra are fitted using Eq. 8.6, where $\delta\omega$ is given by 8.4 and $P(\Omega)$ in this case represents the distribution of the principal director in the angle θ in the plane of rotation of the sample (the sample rotates around an axis perpendicular to $\mathbf{B}_0$ as shown in Fig. 8.8).

The fitting parameters are compatible with the asymmetry parameter $\eta = 0$, which means that the results presented in the figure are fully compatible with a uniaxial nematic phase, both for the substituted and nonsubstituted monomers. The quality of the theoretical fits improves if a nonuniform distribution of the director **n** in θ is considered. Nevertheless, the essential of the uniaxial

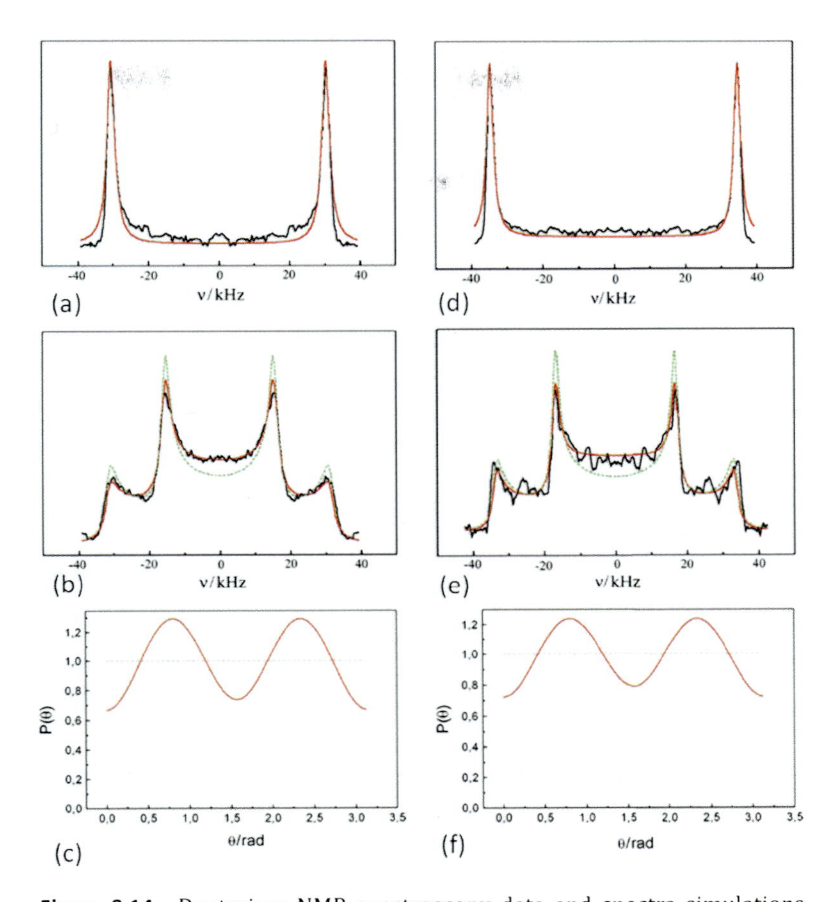

Figure 8.14 Deuterium NMR spectroscopy data and spectra simulations for the monomers associated with organosiloxane tetrapodes mixed with deuterated probe 7CBαd2 presented in Fig. 8.3: (a) Spectrum of aligned sample of substituted monomer; (b) spectrum of rotating sample of substituted monomer; (c) director radial distribution of substituted monomer; (d) spectrum of aligned sample of nonsubstituted monomer;(e) spectrum of rotating sample of nonsubstituted monomer; and (f) director radial distribution of nonsubstituted monomer. In (a, b, d, e) Black line, experimental data; green line, fit with uniform 2D powder; red line, fit with nonuniform 2D powder with director distributions (c and f). From [Cardoso, M., Figueirinhas, J. L., Cruz, C., Van-Quynh, A., Ribeiro, A. C., Feio, G., Apreutesei, D. and Mehl, G. H. (2008). Deuterium NMR investigation of the influence of molecular structure on the biaxial ordering of organosiloxane tetrapodes nematic phase, *Molecular Crystals and Liquid Crystals* **495**, pp. 348–359] reprinted by permission of Taylor & Francis Ltd, www.tandfonline.com.

signature of the phase remains: the splitting of the inner peaks is equal to 1/2 of the splitting of the external ones as shown in Fig. 8.9.

The nonuniform distribution of the director in the angle θ (angle between the director and the NMR static magnetic field $\mathbf{B}_0$) is explained by the stability conditions of molecular alignment associated with the magnetic torque (Cardoso et al., 2008).

This referred work shows that the monomers, which have a molecular structure completely similar to the tetrapode arms, present very different physical properties, namely in terms of molecular ordering. These investigations indicate that the molecular structure resulting from the linkage of the monomers, forming a more complex chemical entity, induces the correlation between the transverse orientations of the mesogenic units giving rise to the biaxial nematic behavior.

The peculiar molecular packing of the tetrapodes' mesophases has also a significant effect on the molecular dynamics. This will be considered in more detail in the next chapter regarding NMR relaxometry studies (Filip et al., 2010).

Chapter 9

NMR Relaxometry of Liquid Crystal Dendrimers: Experimental Results

9.1 Liquid-Crystalline Dendrimers Investigated by NMR Relaxometry

The investigation of molecular movements is of utmost importance for the characterization of liquid-crystalline (LC) dendrimers both in their fundamental and technological oriented aspects. In this chapter we will discuss the application of experimental techniques presented in Chapter 7 to the study of molecular dynamics in the mesophases of various LC dendrimers among those introduced in Chapter 3. As pointed out before, different types of molecular movements exhibited by liquid crystals (namely rotations/reorientations, self-diffusion, and collective modes) can be detected by proton nuclear magnetic resonance (NMR) relaxometry. The characteristic timescales of specific molecular motions vary over several orders of magnitude and can be probed by using values of NMR static field, corresponding to Larmor frequencies of the order of the inverse of those timescales. Therefore, having access to a wide range of NMR frequencies is ideal for molecular dynamics investigations. This is achieved through the combination of standard and fast-

NMR of Liquid Crystal Dendrimers
Carlos R. Cruz, João L. Figueirinhas, and Pedro J. Sebastião

ISBN 978-981-4745-72-7 (Hardcover), 978-981-4745-73-4 (eBook)
www.panstanford.com

field-cycling (FFC) NMR techniques. Typically, in low-molecular-weight liquid crystals, the molecular dynamics timescales vary from picoseconds to nanoseconds for molecular rotations/reorientations, from nanoseconds to microseconds for molecular self-diffusion and from microseconds to milliseconds or slower in the case of collective motions.

As is generally considered in the case of LC systems, the spin-lattice relaxation rate T_1^{-1} introduced in Chapter 6 is given by the sum of the contributions corresponding to different types of molecular motions.

$$T_1^{-1} = \sum_{i=1}^{N} \left(T_1^{-1}\right)_i \tag{9.1}$$

This assumption is valid considering that the molecular movements accessed by NMR relaxometry in the systems under investigation occur within significantly different characteristic timescales.

Typically, the index i in Eq. 9.1 may be associated with molecular rotations/reorientations ($i = R$), molecular self-diffusion ($i = SD$), or collective motions ($i = CM$). Additional contributions may appear in cases where particular relaxation mechanisms take place (e.g., the *cross-relaxation* process that will be mentioned ahead) and some may be absent or have negligible effect due to the structure of the molecules and/or the mesophases. The most obvious case is the lack of collective movements in isotropic phases due to the absence of long-range positional correlations between molecules.

Molecular rotations/reorientations and molecular self-diffusion are thermally activated movements (Eq. 6.40), which give contributions to the relaxation rate dependent on the corresponding characteristic times (correlation times) expressed by Eqs. 6.37, 6.38, 6.39, and 6.47, respectively. Strictly speaking, Eqs. 6.37, 6.38, and 6.39 are obtained for isotropic systems, however, in many cases, they may be applied to mesophases, as good approximations, having in mind a careful interpretation of the physical parameters involved in the problem. This type of approach was followed in some studies of LC dendrimers and the details will be presented in the next subsections. The characteristic time associated to molecular self-diffusion movements (typically, the average time between consecutive diffusive jumps, τ_D, in random jump models, generally applied to liquid crystals) is related to the average square distance of the jump $< r^2 >$ (associated with the average distance between molecules) through the diffusion coefficient D by the expression derived for the isotropic random walk of a particle in three dimensions (Torrey, 1953):

$$D = \frac{< r^2 >}{6\tau_D}. \tag{9.2}$$

In the more complex case of anisotropic molecular self-diffusive movements Eq. 9.2 must be adapted in order to take into account the anisotropy of the system (Žumer and Vilfan, 1978). For instance, in the case of smectic phases, the diffusion coefficient associated to interlayer movements, parallel to the layers normal (perpendicular to the layers) is given by

$$D_{\parallel}^{0} = \frac{< r_{\parallel}^{2} >}{2\tau_{\parallel}} \tag{9.3}$$

and that corresponding to the intralayer displacements (perpendicular to the layer normal) is

$$D_{\perp}^{0} = \frac{< r_{\perp}^{2} >}{4\tau_{\perp}} \tag{9.4}$$

where $\tau_{\parallel}$, $< r_{\parallel}^{2} >$, and $\tau_{\perp}$, $< r_{\perp}^{2} >$ are the average time between consecutive jumps and the average square distance of displacement for diffusion parallel and perpendicular to the director (normal to the layers), respectively (Žumer and Vilfan, 1980).

Those expressions are valid for ideal perfectly aligned phases. The real diffusion coefficients can be obtained from those of ideal completely aligned phases, as a function of the nematic order parameter S, using the expressions (Blinc et al., 1974; Zupancic et al., 1974)

$$D_{\parallel} = D(1 - S) + D_{\parallel}^{0} S \tag{9.5}$$

and

$$D_{\perp} = D(1 - S) + D_{\perp}^{0} S \tag{9.6}$$

with

$$D = \frac{1}{3}(2D_{\perp}^{0} + D_{\parallel}^{0}) \tag{9.7}$$

Besides these correlation times and related diffusion coefficients, a number of other relevant physical parameters are accessible through a detailed analysis of the relaxation times and in particular T_1 as a function of temperature and Larmor frequency.

The collective movements, which appear specifically in LC phases, show particular signatures in the dependence of the relaxation rates (namely T_1^{-1}) on the Larmor frequency ω. As the most clear evidence, already mentioned in Chapter 6, $T_1^{-1}(\omega)$ is proportional to different powers of $1/\omega$ depending on the mesophase under investigation. For instance, $T_1^{-1}(\omega)$ is proportional to $1/\omega^{1/2}$ in the case of director fluctuations typically observed in nematic phases (see Eq. 6.60) and proportional to $1/\omega$ in layer undulations characteristic of smectics (see Eq. 6.63). $T_1^{-1}(\omega)$ also presents a specific behavior in columnar phases described by the spectral density given by Eq. 6.65.

In general, the influence of collective movements on the relaxation rates is also characterized by maximum and minimum cutoff frequencies. Those parameters depend directly on the shortest and largest wavelengths of the oscillation modes, determined by the material under investigation and the experimental conditions. The lower and upper limits of those wavelengths are imposed by the molecular dimensions and the sample size, respectively. However, those are just extreme conditions and, in reality, other circumstances (e.g., the existence of molecular clusters) impose more limited thresholds. As described in Chapter 6, cutoff frequencies depend on elastic constants and effective viscosities and, therefore, the determination of those frequencies gives access to an experimental evaluation of the viscoelastic properties of the studied mesophases.

LC dendrimers are more complex than low-molecular-weight liquid crystals and this additional complexity reflects on their molecular dynamics behavior. In recent years a systematic investigation on proton NMR relaxation of those systems has been reported in the literature. Those studies focus essentially on side-chain LC dendrimers, since they are the most promising systems with respect to applications as mentioned in Chapter 3. A considerable variety of compounds has been investigated, including both side-on and end-on LC dendrimers with diverse chemical structures and different generations (Filip et al., 2005, 2007, 2010; Van-Quynh et al.,

2005, 2006, 2010). The referred studies intend to provide answers to questions that naturally arise when structural complexity is introduced in the molecular units namely: if the types of movements are similar to those of low-molecular-weight liquid crystals; if and how their typical timescales and relevant physical parameters (like elastic constants, viscosities and correlations lengths) are influenced by the structure of the dendritic molecules and their mesophases. In the following we will present the results of investigations on several of such systems. The molecular structure and the mesophases' sequence of each LC dendrimer compound will be described. The main proton NMR relaxation results will be shown together with the fits of the theoretical models corresponding to the different types of molecular movements introduced in Chapter 6.

Finally, the effects of the molecular architecture and mesophases' structure of the LC dendrimers on their molecular dynamics will be discussed by comparing the results obtained from the diversity of systems under investigation.

9.1.1 End-On Organosiloxane Tetrapodes with Strong Terminal Dipoles

In Fig. 9.1 we present the chemical structure and a schematic view of an organosiloxane tetrapode end-on functionalized with mesogenic units bearing a strong terminal dipole, shortly denoted as T-CN. The phase sequence of this compound, synthesized by Mehl and co-workers at the University of Hull, U.K., is also shown in the figure. It's important to notice that the presence of the terminal dipole induces the formation of partially bilayered smectic phases SmA_d and SmC_d. As mentioned in Chapter 2, these phases appear due to the competition between the tendency of smectic ordering (driven by the effect of microsegregation of different molecular elements: aromatic cores, aliphatic chains, and siloxane spacers) and the antiferroelectric ordering associated with the head-to-head coupling of terminal dipoles belonging to neighboring molecules in adjacent layers. Molecular packing models for the mesophases exhibited by these particular systems were obtained from POM and XRD data and reported in the literature (Filip et al., 2007).

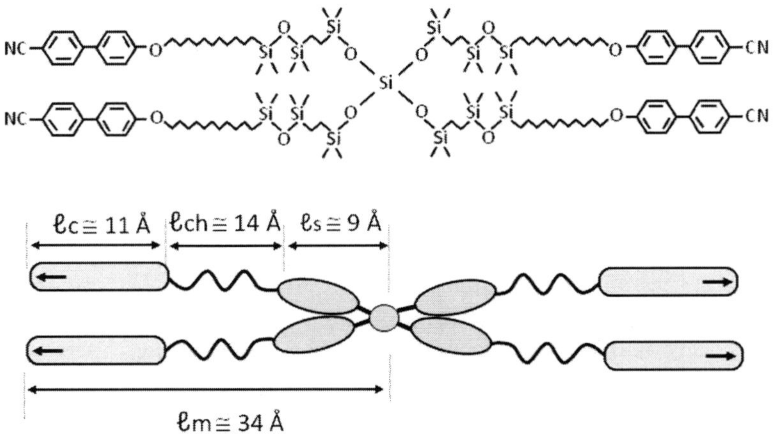

Figure 9.1 Chemical structure, schematic view, and phase sequence of tetrapode T-CN (Filip et al., 2007).

A schematic representation of these packing models is shown in Fig. 9.2.

In the study reported in (Filip et al., 2007), besides the structural characterization of the T-CN tetrapode, molecular dynamics of the system in the isotropic, SmA_d and SmC_d phases, was investigated by means of NMR relaxometry, combining standard and FFC NMR, at Larmor frequencies between 5 kHz and 300 MHz.

The most noticeable feature of the relaxation data in the smectic phases of the compound, presented in Fig. 9.3, is the clear prevalence of a $1/\omega$ contribution, typical of layers undulations, LU, over a wide range of Larmor frequencies. The leveling off of this contribution, to a frequency-independent plateau, occurs for values below 100kHz and can be explained by the low cutoff frequency determined by the largest wavelength of the layer undulation modes. This contribution dominates the relaxation rate until the MHz regime. This trait is not common to low-molecular-weight liquid crystals and its observation is an important contribution of that study (Filip et al., 2007) as will be discussed ahead.

Figure 9.3 shows the plots of the relaxation rate T_1^{-1} as a function of the Larmor frequency at different temperatures corresponding to the isotropic SmA_d, and SmC_d of the compound T_{CN}. The dashed

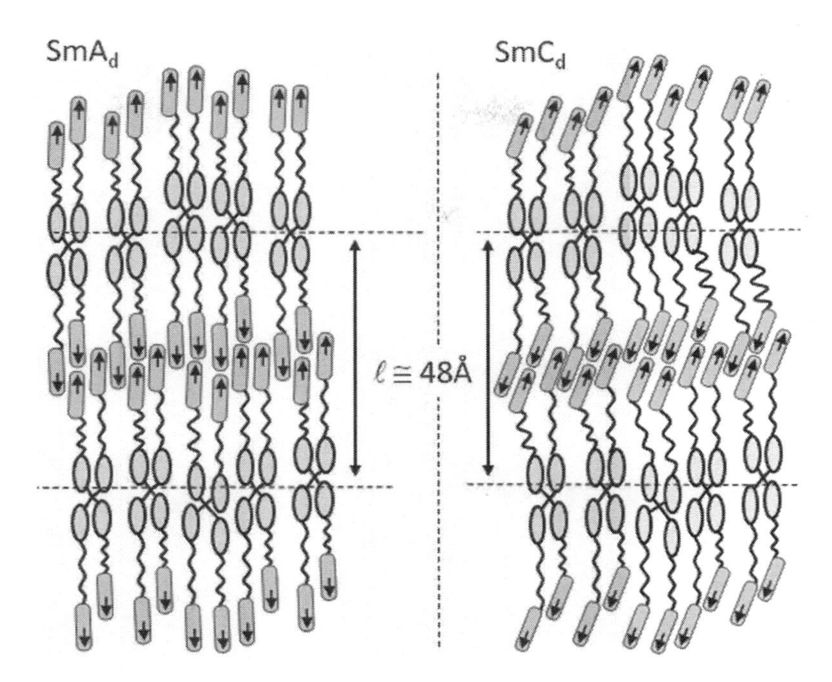

Figure 9.2 Packing model of the SmA$_d$ and SmC$_d$ of organosiloxane tetrapode Ts (Filip et al., 2007).

curves correspond to the theoretical relaxation models presented in Chapter 6, associated with different molecular motions. The solid curve results from the fit of the sum of the partial contributions to the experimental data according to Eq. 9.1.

Besides the layer undulation term, LU, three BPP additional contributions BPP1, BPP2, and BPP3 (see Eq. 6.37), were considered necessary to explain the overall frequency dispersion at higher frequencies.

The overall expression to be fitted to the experimental data may be written as

$$\frac{1}{T_1} = \sum_{i=1}^{3} \left(\frac{1}{T_1}\right)_{\mathrm{BPP}_i} + \left(\frac{1}{T_1}\right)_{LU} \tag{9.8}$$

with (see Eq. 6.37)

$$\left(\frac{1}{T_1}\right)_{\mathrm{BPP}_i} (\omega, A_{ROT_i}, \tau_i) = A_{ROT_i} \left[\frac{\tau_i}{1 + \omega^2 \tau_i^2} + \frac{4\tau_i}{1 + 4\omega^2 \tau_i^2}\right] \tag{9.9}$$

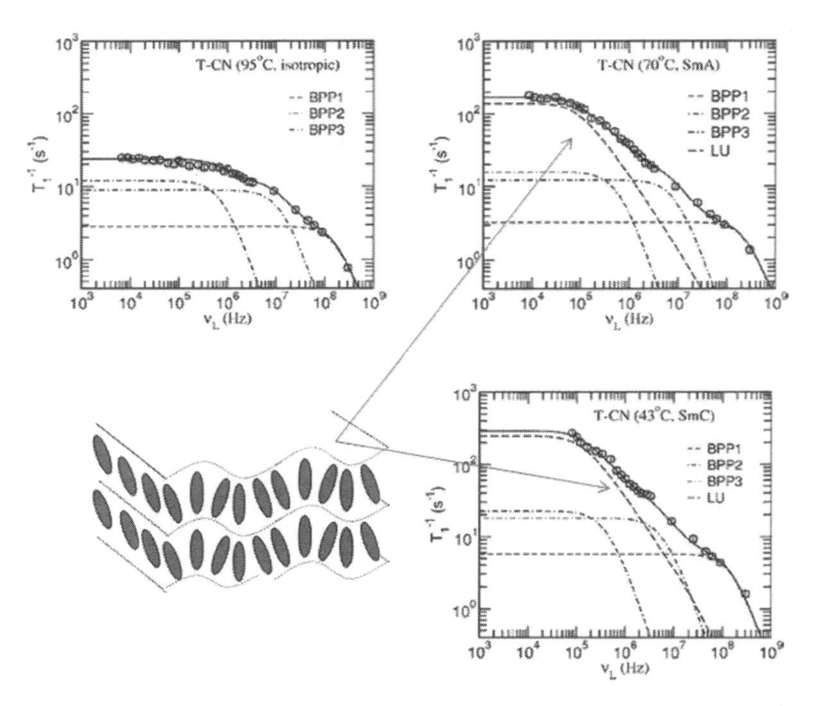

Figure 9.3 Fits of theoretical relaxation models to the experimental T_1^{-1} data as a function of the Larmor frequency for temperatures in the isotropic (top left), SmA_d (top right), and SmC_d (bottom) of the organosiloxane tetrapode T_{CN}. The dashed lines correspond to the contributions of the relaxation mechanisms: three BPP curves, BPP1, BPP2, and BPP3 (see Eq. 6.37), and a contribution of collective motions, in this case associated to layer undulations (LUs) (see Eqs. 6.63 and 6.24). The solid line corresponds to the sum of the contributions to the relaxation rate according to Eq. 9.1. Reprinted (figure) with permission from [Filip, D., Cruz, C., Sebastiao, P. J., Ribeiro, A. C., Vilfan, M., Meyer, T., Kouwer, P. H. J. and Mehl, G. H. (2007). Structure and molecular dynamics of the mesophases exhibited by an organosiloxane tetrapode with strong polar terminal groups, *Physical Review E* **75**, 1, p. 011704]. Copyright (2007) by the American Physical Society.

A_{ROT_i} are the strengths of the BPP contributions, which depend on the spins density, the interproton distances and the number of protons in a molecule, involved in each of the rotational movements considered (see Eqs. 6.22, 6.23, 6.24, and 6.37). τ_i are the correlation times characteristics of each of the partial rotation/reorientation movements.

Considering that the rotation movements are thermally activated, the temperature dependence of each of the τ_i was calculated following an Arrhenius law (see Eq. 6.40):

$$\tau_i = \tau_{i,\infty} e^{E_{ai}/(RT)} \tag{9.10}$$

where $\tau_{i,\infty}$ are the extrapolated values of τ_i in the limit where the temperature T tends to infinity, E_{ai} are the activation energies (J/mol) for each of the partial movements described by the BPP contributions, and R is the universal gas constant.

According to Eq. 6.63 the contribution of layer undulations is given by:

$$\left(\frac{1}{T_1}\right)_{LU} (\omega, A_{LU}, \omega_{\mathrm{cM}}, \omega_{\mathrm{cm}}) = \frac{A_{LU}}{\omega} \left[\arctan\left(\frac{\omega_{\mathrm{cM}}}{\omega}\right) - \arctan\left(\frac{\omega_{\mathrm{cm}}}{\omega}\right)\right] \tag{9.11}$$

where A_{LU} is a constant which depends on the viscoelastic properties of the material, and ω_{cM} and ω_{cm} are the high and low cut-off frequencies of director fluctuations given by $K_1 q_{\mathrm{max}}^2/\eta$ and $K_1 q_{\mathrm{min}}^2/\eta$, respectively. K_1 denotes the splay elastic constant and η the corresponding viscosity. q_{max} and q_{min} are the highest and the lowest wavenumbers of the oscillation modes associated with the layer undulations.

It is worthwhile to notice that the LU expression (Eq. 9.11), considering Eqs. 6.63, 6.28, 6.29, 6.30, 6.31, and 6.32, is valid both for aligned samples (director parallel to the static NMR field B_0) and for any value of the angle (Δ) between the director and B_0, since the only difference between the spectral densities $J_k(\Delta, \omega)$ is a constant $f_{k1}(\Delta)$ independent of the frequency ω (see Eq. 6.63).

The results presented in Fig. 9.3 were analyzed by means of a global nonlinear least-square minimization procedure, in which Eqs. 9.8, 9.9, 9.10, and 9.11 were fitted simultaneously to the experimental data.

In the case of the isotropic phase only the three BPP contributions were considered, since no evidence of collective movements exist in that case, as indicated by the wide *plateau* on the low-frequency region ($\frac{\omega}{2\pi} < 1 MHz$). In that case, the factors A_{BPP_i} and the three correlation times τ_i, were considered as fitting parameters.

In what concerns the smectic phases, also A_{LU}, ω_{cM}, and ω_{cm}, characteristic of the layer undulation contribution, were taken into account.

The three BPP contributions correspond to correlation times, τ_1, τ_2 and τ_3 of the order of 10^{-10}s, 10^{-9} – 10^{-8}s, and 10^{-7}s, respectively. These terms may be assigned to rotational reorientation movements of the dendritic arms and their molecular segments (mesogenic units, siloxane linkers and core). The three different factors A_{BPP1}, A_{BPP2}, and A_{BPP3}, are associated to the number proton pairs in the molecular segment performing the corresponding reorientational motion (Filip et al., 2007). The relatively small values of the activation energies $E_{a1} \approx 19$ kJmol^{-1}, $E_{a2} \approx 13$ kJmol^{-1}, and $E_{a3} \approx 12$ kJmol^{-1} (see Eq. 9.10) (comparing with known values of low-molecular-weight liquid crystals) are compatible with the assumption of the partial reorientation movements of the dendritic arms and molecular segments.

As in other LC phases of dendrimers (Filip et al., 2010; Van-Quynh et al., 2005, 2006, 2010) the contribution of a global molecular self-diffusion process of the dendritic molecules to the proton relaxation rate is not found in the analysis of the experimental data. This effect is clearly indicated by the absence of the plateau, between 100 kHz and a few MHz, typical of this type of movements.

This is not unexpected, since the highly probable interdigitation of dendritic arms belonging to different neighboring dendrimers will make the self-diffusion of a whole dendrimer a very slow process. However, self-diffusion of a single dendrimer, mediated by sequential lateral displacements of the dendritic arms within a layer is not excluded. Interlayer diffusion of the whole dendrimers is expected to be even more unlikely (ant therefore even slower) due to the microsegregation between aromatic rigid cores, aliphatic chains, and siloxane dendritic cores in separated sublayers combined with the covalent linkage of the mesogenic units to the central siloxane cores (see Fig. 9.2). The slowing down of molecular self-diffusion, and the consequent lack of the corresponding middle frequencies plateau, contributes to a clearer identification of the LU mechanism, which, in this case, dominates the frequency dispersion curves up to 1 MHz. The value of A_{LU} determined from the fits ($\sim 10^7 s^{-2}$) is significantly higher than typical values found in low-molecular-weight liquid crystals (e.g., $\sim 10^4 s^{-2}$ in the compound DB_8Cl) (Filip et al., 2007). A low correlation length (in the direction perpendicular to the layers) of the layer undulations was appointed in (Filip et al., 2007) as a plausible explanation for this result. This correlation length ξ can be estimated taking into account the expression of A_{LU} obtained from Eqs. 6.9, 6.24, 6.45, and 6.63

$$A_{LU} = \left(\frac{\mu_0}{4\pi}\right)^2 \frac{9\gamma^4\hbar^2 k_B T S^2}{64\pi K_1 \xi} \left\langle \frac{\left[3\cos^2(\alpha_{ij}) - 1\right]^2}{r_{ij}^6} \right\rangle \tag{9.12}$$

where r_{ij} is the interproton distance of a spin pair ij, α_{ij} is the angle between the proton pair vector and the long molecular axis (see Eq. 6.45), S is the orientational order parameter, and K_1 is the splay elastic constant of the mesogenic sublayer. Assuming typical values for the physical parameters in Eq. 9.12 found for low-molecular-weight thermotropic liquid crystals ($r_{ij} \approx 1.8$ Å, $S \approx 0.7$, and $K_1 \approx 10^{-11}$ N) we obtain $\xi = 58$ Å for the correlation length. This value is in very good agreement with the layer thickness $\ell = 48$ Å, as shown in Fig. 9.2 (Filip et al., 2007).

Such a short correlation length for layer undulations might appear contradictory with the (quasi) long-range positional ordering as observed by X-ray diffraction (XRD) in the smectic phases of T-CN. However, it is not necessarily so. The XRD gives a picture of the (average) static molecular positional correlations whilst the NMR relaxation results are related to the dynamic characteristics of the system. The short correlation length measured by NMR indicates that the individual smectic layers oscillate almost independently of each other, although keeping very well-defined positional correlations (as shown by X-rays). Actually, this is not surprising taking into account the separation between smectic layers defined by the microsegregation of the molecular segments and their linkage to the central siloxane cores. Such a phase structure induces the decoupling between the oscillations of the individual layers. It is worthwhile to notice that this type of behavior is common in lyotropic liquid crystals (Halle, 1994; Vilfan et al., 2001).

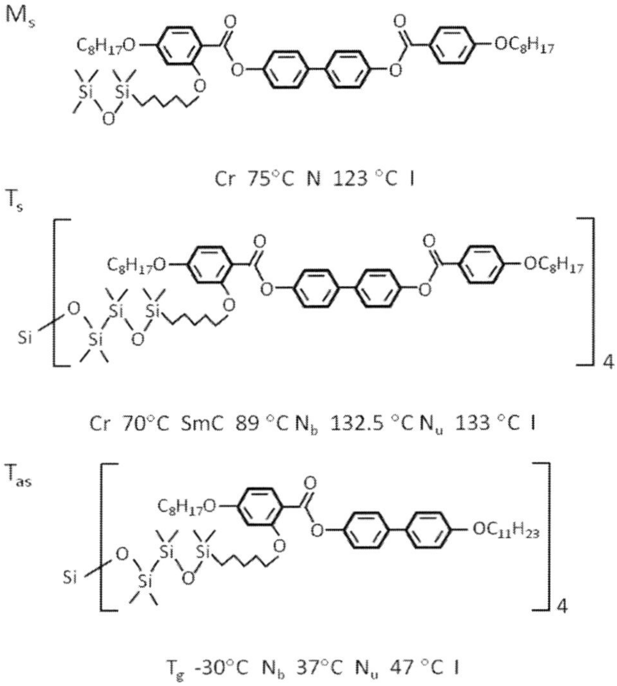

M_s

Cr 75°C N 123 °C I

T_s

Cr 70°C SmC 89 °C N_b 132.5 °C N_u 133 °C I

T_{as}

T_g -30°C N_b 37°C N_u 47 °C I

Figure 9.4 Chemical structure and phase sequence of monomer Ms (top) and organosiloxane tetrapodes Ts (middle) and Tas (bottom).

9.1.2 Organosiloxane Tetrapodes with Laterally Attached Mesogens

Side-on organosiloxane tetrapodes (generation 0 dendrimers) with (i) four benzene rings (Ts: symmetric) and (ii) three benzene rings, (Tas: asymmetric) have been studied by proton NMR relaxation over a wide Larmor frequency range, and for temperatures corresponding to the different mesophases exhibited by the materials. The monomer corresponding to the tetrapode with symmetric mesogen, Ms, was also studied using the same technique in order to compare the effect of branching on the molecular dynamics. The chemical structures and phase sequences of compounds Ms, Ts, and Tas are shown in Fig. 9.4. These compounds were synthesized by Mehl and co-workers at the University of Hull, U.K. (Filip et al., 2010).

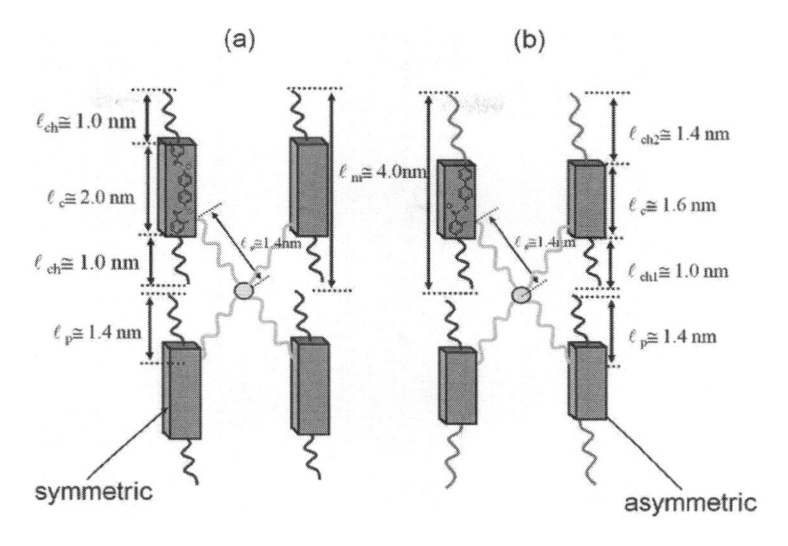

Figure 9.5 Schematic view of organosiloxane tetrapodes Ts and Tas.

It is important to notice that the difference between the chemical structure of the monomer Ms and the corresponding tetrapode Ts is uniquely the connection of the mesogenic units to the central linking group. Nevertheless, the effect of the branching on the mesomorphic behavior of the materials is crucial as shown by the phases sequences presented in Fig. 9.4. First of all, the smectic C phase exhibited by the tetrapode doesn't exist in the monomer Ms phases' sequence and, on the other hand, the mesophases of the tetrapode appear at much lower temperatures. As described in the literature (Filip et al., 2010), linking the four mesogenic monomers in this particular dendrimer clearly stabilizes the lamellar phase (SmC) by favoring the microsegregation between the aromatic cores in a sublayer and the aliphatic plus siloxane chains on another one.

In Fig. 9.5 we present a schematic view of both Ts and Tas tetrapodes, including the mesogenic cores the aliphatic chains and the siloxane spacers which connect the monomers to the central tetravalent silicon atom.

The packing models of the nematic and SmC phases exhibited by those materials, resulting from polarizing optical microscopy (POM) and XRD investigations are presented in Fig. 9.6 (Filip

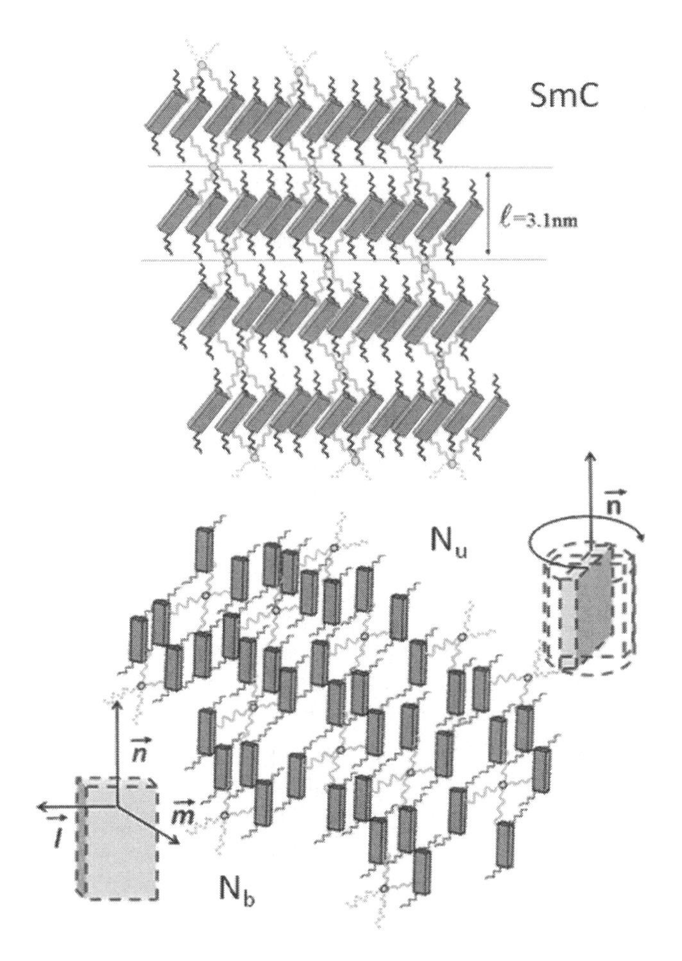

Figure 9.6 Packing model of nematic and smectic C phases of organosiloxane tetrapode Ts.

et al., 2010). As discussed in detail by Filip et al. (2010), the layers in the SmC phase are formed by interdigitated mesogenic units belonging to neighboring tetrapodes, eventually belonging to adjacent layers. The dendritic central cores (tetravalent silicon atoms) are microsegregated remaining, in average, in planes separating two contiguous layers. In average, for each tetrapode, two mesogenic units stay in the "upper" layer and the remaining two in the "lower" one. In the case of the nematic phase, similar molecular packing remains at the local level (cybotatic clustering) but the overall layered structure is destabilized by the higher temperature of that phase. Interestingly, X-ray data show beyond any doubt that, contrary to what is commonly observed in low-molecular-weight liquid crystals (where the smectic cybotactic clusters exist only for temperatures close to the smectic-nematic phase transition temperature), the cybotactic structure exhibited by these tetrapodes are present over the whole nematic temperature range and, eventually, in the isotropic phase, close to the N-I phase transition. As we will see next, these results are consistent with the proton NMR relaxation data.

In Fig. 9.7 we present the results of spin–lattice relaxation rates (inverse of relaxation time T_1) over several decades of the Larmor frequency for the isotropic and nematic phases of monomer Ms and tetrapode Tas. We are comparing specifically Ms with Tas (and not with the tetrapode Ts which has branches similar to that monomer) because Tas exhibits a similar phases' sequence with a single nematic phase. However, the conclusions regarding the nematic phase of Tas are similar to those corresponding to Ts. The dashed curves in the figure correspond to the contributions of the different types of molecular motions discussed in Chapter 6 to the relaxation rate, the solid curve represents the fit of sum of those contributions to the experimental data according to Eq. 9.1.

In general, as described previously (see Eq. 9.1), the expression for the proton spin–lattice relaxation rate of LC systems includes the sum of contributions from molecular reorientations/rotations, self-diffusion, and collective motions. The T_1^{-1} frequency dispersion corresponding to the Ms system (monomer) is similar to the usually observed in low-molecular-weight liquid crystals. The results in the isotropic phase are very well explained by the sum of an SD contribution described by the Torrey model (see Eqs. 6.24 and 6.47) and a BPP contribution associated to molecular rotations (Eq. 6.37). In this case the relaxation rate is given by Eq. 9.13:

$$\frac{1}{T_1} = \left(\frac{1}{T_1}\right)_{\mathrm{BPP}} + \left(\frac{1}{T_1}\right)_{\mathrm{SD}} \tag{9.13}$$

$$\frac{1}{T_1} = \left(\frac{1}{T_1}\right)_{\mathrm{BPP}} + \left(\frac{1}{T_1}\right)_{\mathrm{SD}} + \left(\frac{1}{T_1}\right)_{\mathrm{ODF}} \tag{9.14}$$

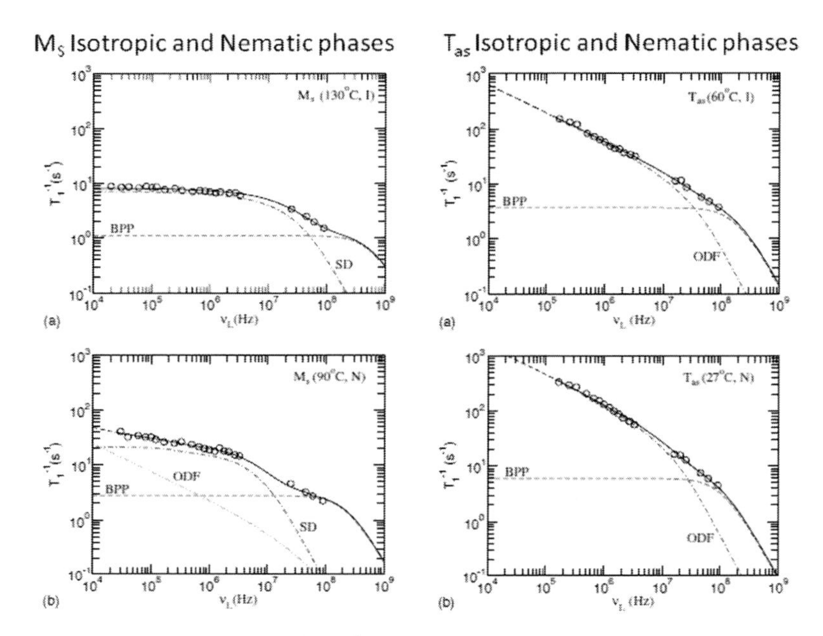

Figure 9.7 Relaxation rate (T_1^{-1}) as a function of the Larmor frequency and fits of relaxation models for the isotropic and nematic phases of monomer Ms and organosiloxane tetrapode Tas. Reprinted (figure) with permission from [Filip, D., Cruz, C., Sebastiao, P. J., Cardoso, M., Ribeiro, A. C., Vilfan, M., Meyer, T., Kouwer, P. H. J. and Mehl, G. H. (2010). Phase structure and molecular dynamics of liquid-crystalline side-on organosiloxane tetrapodes, *Physical Review E* **81**, 1, p. 011702]. Copyright (2010) by the American Physical Society.

In the nematic phase the inclusion of an additional contribution, proportional to $\omega^{-1/2}$ is necessary to fit the theoretical curve of the proton relaxation rate to the experimental data (see Eq. 9.14). That term, given by Eqs. 6.24, 6.60, and 6.61 is typical of order director fluctuations (ODFs), the type of collective movements generally observed in nematics. It may be written as:

$$\left(\frac{1}{T_1}\right)_{\mathrm{ODF}}(\omega, A_{\mathrm{ODF}}, \omega_{\mathrm{cM}}, \omega_{\mathrm{cm}}) = \frac{A_{\mathrm{ODF}}}{\omega^{1/2}}\left[f\left(\frac{\omega_{\mathrm{cM}}}{\omega}\right) - f\left(\frac{\omega_{\mathrm{cm}}}{\omega}\right)\right] \tag{9.15}$$

where

$$f(x) = \frac{1}{\pi}\left[\arctan(\sqrt{2x}+1) + \arctan(\sqrt{2x}-1) - \operatorname{arctanh}\left(\frac{\sqrt{2x}}{1+x}\right)\right] \quad (9.16)$$

A_{ODF} is a term that depends on the temperature, on an effective viscosity, η and on an elastic constant K (in a one constant approximation). $f(x)$ is the cutoff function depending on the high and low cutoff frequencies

$$\begin{aligned}\omega_{\mathrm{cM}} &= Kq_{\mathrm{max}}^2/\eta \\ \omega_{\mathrm{cm}} &= Kq_{\mathrm{min}}^2/\eta\end{aligned} \quad (9.17)$$

where q_{max} and q_{min} are the highest and the lowest wave number of the fluctuation modes, respectively.

In the case of the monomer Ms, the values of the models' parameters (the strength of the BPP contribution, A_{BPP} (Eq. 6.37), the correlation times characteristics of rotational reorientations, τ_R, and molecular self-diffusion τ_{D}, the spins density, n, and closest distance between two molecules, d, the strength of the ODF fluctuations and related factors are compatible to those found in low-molecular-weight liquid crystals (Filip et al., 2007).

Something quite different happens in the case of the dendrimer Tas (and with Ts, as will be shown ahead). Considering the dependence of the spin–lattice relaxation rate on the temperature for Tas, presented in Fig. 9.7, it is evident that, contrary to what is verified for the monomer, the $T_1^{-1}(\omega)$ curve shows an almost monotonous increase with decreasing frequency for $\omega < 100$ MHz, both in the nematic and isotropic phases. This behavior is not usually seen in low-molecular-weight liquid crystals. In the case of Tas, the lack of the intermediate frequencies plateau, referred before, is compatible with the absence of the molecular self-diffusion influence on the relaxation rate. Taking into account the packing model deduced from XRD data, molecular self-diffusion is strongly hindered due to the interdigitation of meseogenic units resulting in a somehow imbricated phase structure (see Fig. 9.6). On the other hand, the almost monotonous variation of the relaxation rate can be attributed to a $T_1^{-1}(\omega)$ proportional to $\omega^{-1/2}$ dependence, typical of ODFs in nematic phases (see Eq. 9.15). In this case the overall relaxation rate is given just by the sum of two terms (see Fig. 9.7):

$$\frac{1}{T_1} = \left(\frac{1}{T_1}\right)_{\mathrm{BPP}} + \left(\frac{1}{T_1}\right)_{\mathrm{ODF}} \tag{9.18}$$

More surprising is the observation of the same type of behavior (see Eq. 9.18) and Fig. 9.7 in the isotropic phase of Tas where the existence of ODFs is unexpected. As reported in (Filip et al., 2005) this result can be explained by the subsistence of nematic cybotactic clusters in the isotropic phase. This local order effect, which was confirmed later by XRD measurements (Filip et al., 2010), is due to the prevalence of molecular interdigitation between neighboring dendrimers in the isotropic phase.

Complete tables and further details of the discussion on the actual values of the fitting parameters for the compounds Ms, Ts, and Tas can be consulted in the work by Filip et al. (2010).

In Fig. 9.8 we present the T_1^{-1} data as a function of the Larmor frequency, for different temperatures corresponding to the isotropic, nematic and smectic C phases of tetrapode with side-on mesogens with four benzene rings (Ts). The solid curve represent the fit of the sum of theoretical relaxation rates presented in Chapter 6 to the experimental data according to Eq. 9.1. The dashed lines represent the contributions of the different molecular motions to the overall relaxation rate.

The nematic phase of Ts show a behavior similar to Tas with respect of the frequency dependence of the relaxation rate T_1^{-1} with a BPP contribution resulting from molecular rotational reorientations and the $1/\omega^{1/2}$ dependence characteristic of nematic ODFs (see Eq. 9.18). However, in the case of the isotropic phase of Ts, an additional contribution, assigned to molecular self-diffusion must be considered to explain the experimental data (Eq. 9.14). As in the case of Ms, this term can be described by the Torrey model for the relaxation induced by molecular self-diffusion, using Eqs. 6.24 and 6.47 (Torrey, 1953). Such a term can be due to diffusive movements of the dendrimer, driven by successive displacements of the dendritic arms. Taking into account that molecular self-diffusion is a thermally activated process, one possible explanation for the

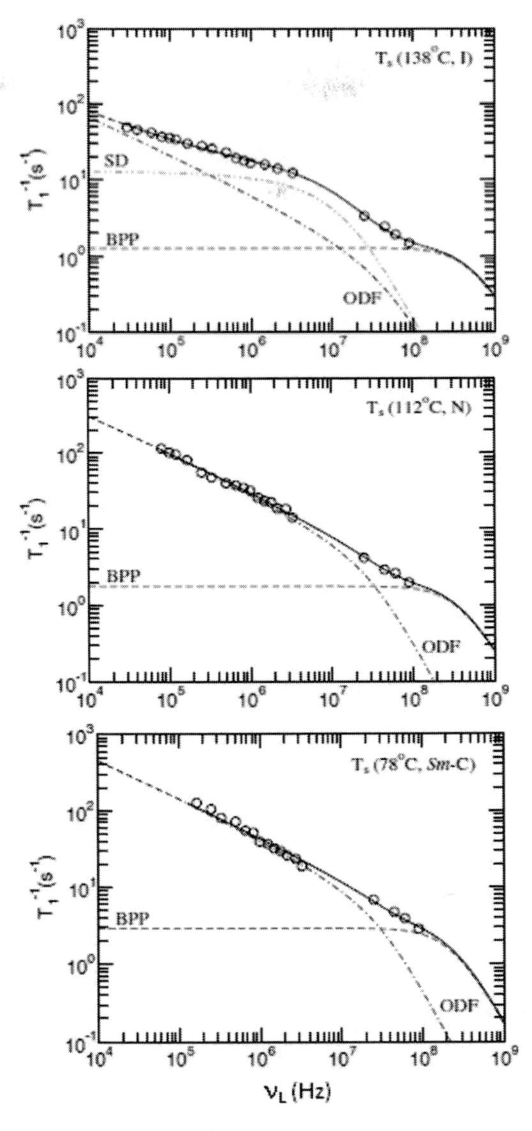

Figure 9.8 Relaxation rate (T_1^{-1}) as a function of the Larmor frequency and fits of relaxation models for the isotropic, nematic and smectic C phases of organosiloxane tetrapode Ts. Reprinted (figure) with permission from [Filip, D., Cruz, C., Sebastiao, P. J., Cardoso, M., Ribeiro, A. C., Vilfan, M., Meyer, T., Kouwer, P. H. J. and Mehl, G. H. (2010). Phase structure and molecular dynamics of liquid-crystalline side-on organosiloxane tetrapodes, *Physical Review E* **81**, 1, p. 011702]. Copyright (2010) by the American Physical Society.

appearance of this contribution, in the case of Ts (and not in Tas), is the higher temperature range of isotropic phase of this compound (compared to Tas, see Fig. 9.4) (Filip et al., 2010).

Remarkably, the frequency dependence of the relaxation rate in the SmC phase of the dendrimer Ts, presents a very similar profile to that of the nematic phases following strictly a $1/\omega^{1/2}$ over a large frequency range (see Figs. 9.8 and 9.7). The fits presented in Fig. 9.8 are given by Eq. 9.18. As in the previous case, the fitting parameters are the BPP strength, A_{BPP} and the respective correlation time τ_R for the molecular rotation mechanism (see Eq. 9.9) and the strength of the ODF mechanism A_{ODF} and high and low cutoff frequencies ω_{cM} and ω_{cm}, respectively (Eq. 9.15).

A contribution proportional to $1/\omega^{1/2}$ is, in some cases, found in the SmC phases, in addition to the $1/\omega$ term, characteristic of smectic layer undulations. That contribution, typical of ODFs in nematics, is due to the in-plane oscillations of the so called *c director*, which correspond to the preferential orientation of the long molecular axis projection in the plane of the (disordered) layers. However, the relaxation rate frequency dependence is quite peculiar, both in Ts and Tas, due to the absence of the layer undulation contribution. This is completely different from what was found in the tetrapode T-CN, with terminally attached mesogens, where layer undulations were the dominant relaxation mechanism over a large frequency range. Contrary to the T-CN case, where the layer movements were mutually decoupled, in the SmC phase of Ts, the tetrapodes are strongly interconnected due to the systematic interdigitation of mesogenic units belonging to adjacent layers (see Fig. 9.6). That interpenetration of the dendrimers imposes severe restrictions to the independent movement of layers resulting in the dumping of the LU mechanism.

The similarity between the collective movements in the SmC and Nematic phases of Ts is explained in (Filip et al., 2010) by the similarity of the molecular packing in those phases. As shown by XRD data, SmC-like clusters survive over all the temperature range of the nematic phases of this compound. The same type of molecular packing is observed in the tetrapode Tas (Filip et al., 2010). This kind of cybotactic behavior is related with the emergence of biaxial ordering (N_B) for lower temperatures, below the SmC-N_B phase transition (Cruz et al., 2008; Figueirinhas et al., 2005).

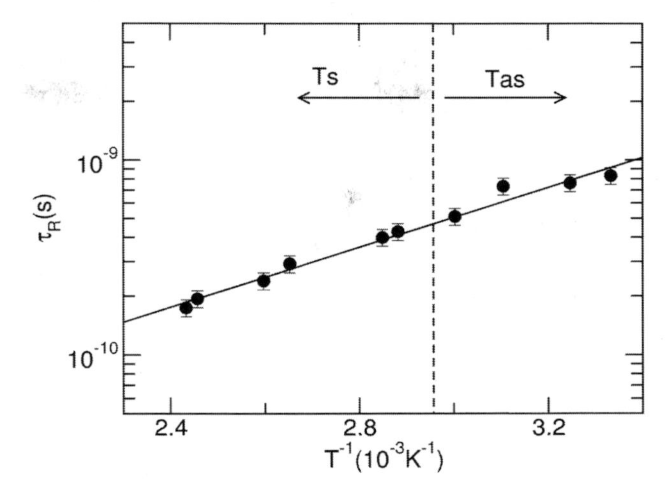

Figure 9.9 Inverse temperature dependence of τ_R in Ts and Tas (isotropic, nematic, and smectic C phases) and fit with Arrhenius law (Eq. 9.10): $\tau_\infty \approx 2.5 \times 10^{-12}$ s, activation energy $E_a = 15$ kJ/mol. Reprinted (figure) with permission from [Filip, D., Cruz, C., Sebastiao, P. J., Cardoso, M., Ribeiro, A. C., Vilfan, M., Meyer, T., Kouwer, P. H. J. and Mehl, G. H. (2010). Phase structure and molecular dynamics of liquid-crystalline side-on organosiloxane tetrapodes, *Physical Review E* **81**, 1, p. 011702]. Copyright (2010) by the American Physical Society.

With respect to the high-frequency domain ($\omega/2\pi > 30$–50 MHz) the frequency dispersion of the relaxation rate is well described by a BPP contribution both in Ts and Tas, and also in the case of the monomer Ms. The fits of the relaxation models to the experimental data are compatible with a similar prefactor A_{BPP} independent of the temperature in all mesophases for the monomer and both tetrapodes. This is interesting and clearly compatible with the assumption that the BPP contribution corresponds to the reorientation movements of the mesogenic units and is mainly influenced by the local molecular arrangement and not by the long distance phase correlations. It is worthwhile to recall that the monomer Ms has exactly the same chemical structure as the mesogenic units of the tetrapode Ts and that the monomers of Tas are closely similar to those of Ts and present the same dimensions in their most extended configurations as schematically depicted in Fig. 9.5.

Moreover, if we plot the correlation time τ_R, characteristic of molecular reorientations, as a function of the inverse temperature both for Ts and Tas over the whole measured temperature range (including all the phases) we verify that the complete set of data is compatible with the same Arrhenius law (see Eq. 9.10) with $\tau_\infty \approx 2.5 \times 10^{-12}$ s and the same activation energy $E_a = 15$ kJ/mol.

In summary, the experimental results show that rotational reorientation movements of the dendrimers' laterally attached mesogenic units are independent of the mesophases, as detected by NMR relaxation in the compounds Ts and Tas. Similar results for high frequencies are also found in the monomer Ms, compatible to those usually obtained in low-molecular-weight liquid crystals with correlation times τ_R of the order of 10^{-10} s. For frequencies below 10 MHz, the molecular dynamics behavior of the dendrimers is quite different from that of the monomer. The molecular self-diffusion mechanism usually found in low-molecular-weight liquid crystals (as in the case of Ms with characteristic times τ_D of the order of 10^{-8}s) is not observed in the tetrapodes confirming the hypothesis of severe molecular mobility restrictions imposed by the interdigitation of mesogenic units belonging to neighboring dendrimers. In this large frequency range (100 kHz $< \omega/2\pi <$ 10 MHz) the relaxation rate is dominated by a contribution proportional to $\omega^{-1/2}$ characteristic of nematic like ODFs. This is valid for the nematic and SmC mesophases and also in the isotropic phase, in that case due to the existence of local order with nematic like oscillation modes, also for temperatures above the N-I phase transition.

9.1.3 Organosiloxane Octopodes with Laterally Attached Mesogens

As described in Chapter 3, Section 3.2.2.2, siloxane-based LC dendrimers can be obtained using a cubic silsexquioxane molecular structure, appropriately functionalized at the eight vertices, as central dendritic core. In a study reported by Sebastião and co-workers, a side-on generation zero dendrimer (octopode) synthesized at the University of Hull by the group of G. H. Mehl was investigated by proton NMR relaxation (Van-Quynh et al., 2010). The central cubic

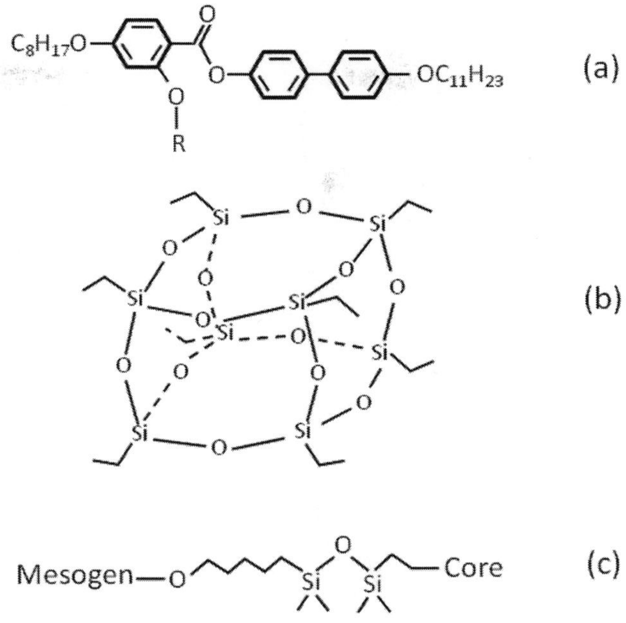

(a)

(b)

(c)

Iso 57 °C N 44 °C Col_h

Figure 9.10 Molecular structure and phase sequence of the organosiloxane octopode studied by means of NMR relaxometry: (a) side-on mesogenic unit, (b) silsexquioxane cubic dendritic core, and (c) siloxane spacer (Van-Quynh et al., 2010).

molecular core of the octopode bears at each of the eight vertices a mesogenic unit similar to the one used in tetrapode Tas, linked by a similar siloxane spacer, see Fig. 9.4. The molecular structure of octopode compound and the corresponding phases' sequence is shown in Fig. 9.10.

This compound exhibits a nematic phase between 57°C and 44°C and a columnar hexagonal phase for temperatures lower than 44°C. As mentioned before (Section 3.2.2.2), for temperatures below the nematic-columnar transition point, the silsexquioxane cubic cages and the mesogenic units are microsegregated in such a way that the silsexquioxane are arranged sequentially along the columnar axes, surrounded by a disordered media composed by the mesogenic

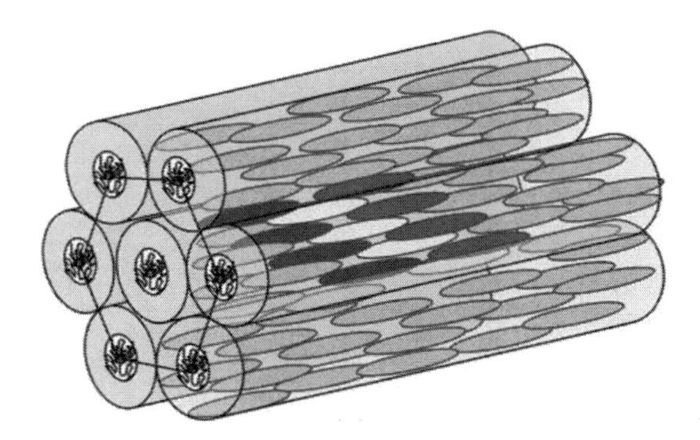

Figure 9.11 Molecular packing of the hexagonal columnar phase of organosiloxane octopodes. Reproduced from Karahaliou, P. K., Kouwer, P. H. J., Meyer, T., Mehl, G. H. and Photinos, D. J. (2007). Columnar phase structures of an organic-inorganic hybrid functionalized with eight calamitic mesogens, *Soft Matter* **3**, 7, pp. 857–865, with permission of The Royal Society of Chemistry.

units (laterally linked by the siloxane spacers to the vertices of the cubes) with their long axes oriented parallel to the columnar axes direction (Karahaliou et al., 2007). See Fig. 9.11.

In Fig. 9.12 we present the results of the spin–lattice relaxation time T_1 as a function of the Larmor frequency for temperatures in the isotropic, nematic, and columnar hexagonal phases of the organosiloxane octopode with side-on mesogenic units. The solid curves represent the fit of the sum of theoretical relaxation rates (T_1^{-1}) presented in Chapter 6 to the experimental data according to Eq. 9.1. The dashed lines represent the contributions of the different molecular motions to the overall relaxation rate.

Both in the isotropic and nematic phases of the octopode, the relaxation rate dependence on the frequency presents a similar behavior to that found in the tetrapode Tas. In that case T_1^{-1} can be described by a sum of a BPP function that accounts for the fast molecular rotations/reorientations (with a correlation times $\tau_1 \approx 10^{-10}$s) with a term associated with nematic ODFs (see Eqs. 9.18, 9.15, 9.16, and 9.9). This is not surprising since, as shown by the XRD results, the molecular organization of the octopode phases has some similarity with those of the Tas tetrapode at local level (Filip et al., 2010; Karahaliou et al., 2007).

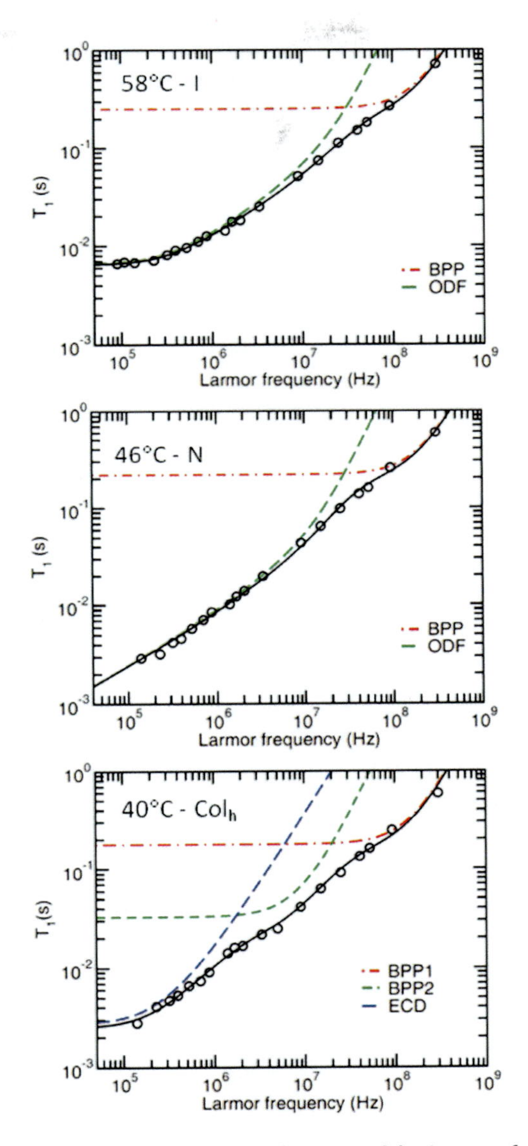

Figure 9.12 Relaxation time (T_1) as a function of the Larmor frequency and fits of relaxation models for temperatures corresponding to the isotropic (56°C), nematic (46°C), and columnar hexagonal (40°C) phases of the studied organosiloxane octopode. From Van-Quynh, A., Sebastiao, P. J., Wilson, D. A. and Mehl, G. H. (2010). Detecting columnar deformations in a supermesogenic octapode by proton NMR relaxometry, *European Physical Journal E* **31**, 3, pp. 275–283. With permission of Springer.

It is interesting to verify that the existence of local nematic order is also detected in the isotropic phase exhibited by the octopodes. Again, this is due to the interdigitation of the lateral (side-on) mesogenic units that forces the preservation of some cybotactic nematic order in the isotropic phase (Van-Quynh et al., 2010).

As reported in (Van-Quynh et al., 2010), the theoretical description of the mechanisms contributing to the relaxation rate in the $\mathrm{Col_h}$ phase is different from that corresponding to the nematic phase. In spite of the similar local molecular packing, the long-range positional correlations associated to the columnar phase (see Fig. 9.11) justify the observation of collective movements described as elastic columnar deformations (ECDs). An additional BPP contribution with a characteristic correlation time τ_2 of the order of $\tau_2 \approx 10^{-8}$s, assigned to slower molecular reorientations (probably associated with restrictions imposed by the columnar packing), is also necessary to explain the frequency dependence of the relaxation rate.

The relaxation rate in the columnar phase is, therefore given by:

$$\frac{1}{T_1} = \left(\frac{1}{T_1}\right)_{\mathrm{BPP1}} + \left(\frac{1}{T_1}\right)_{\mathrm{BPP2}} + \left(\frac{1}{T_1}\right)_{ECD} \tag{9.19}$$

with $(1/T_1)_{\mathrm{BPP1}}$ and $(1/T_1)_{\mathrm{BPP2}}$ given by Eq. 9.9 and the term corresponding to elastic columnar deformations as described by Eqs. 6.24 and 6.65.

$$\left(\frac{1}{T_1}\right)_{ECD}(\omega) = \frac{A_{ECD}}{\omega}\int_0^1 \left[\arctan\left(\frac{R\omega_c}{u\omega} + u\frac{\omega_c}{\omega}\right) - \arctan\left(u\frac{\omega_c}{\omega}\right)\right] du \tag{9.20}$$

with

$$\omega_c = \frac{Kq^2_{\parallel\mathrm{high}}}{\eta} \tag{9.21}$$

and

$$R = \frac{Bq^2_{\perp\mathrm{high}}}{Kq^4_{\parallel\mathrm{high}}} \tag{9.22}$$

K and B the elastic constants for bending and compression of the columns, respectively. η is the effective viscosity and $q_{\parallel\mathrm{high}}$ and $q_{\perp\mathrm{high}}$

are the components of the largest wave vector of the deformation parallel and perpendicular to the columns.

As discussed before, with respect to layer undulation (LU), it is convenient to remember that, as well as in other cases of collective movements (ODF and LU), Eq. 9.20 is valid for any value of the angle (Δ) between the director and B_0 (and consequently also for polycrystalline samples) taking into account Eqs. 6.24, 6.28, 6.29, 6.30, 6.31, 6.32, and 6.65.

The details of the data analysis, including the values of the fitting parameters for the different mechanisms (strengths of the BPP and ECD contributions, correlation times τ_1 and τ_2 and cutoff frequencies in the nematic and $\mathrm{Col_h}$ phase, are described by Van-Quynh et al. (2010). One of the results that should be outlined regards the value obtained for the cutoff frequency of the ECD process ω_c. Using Eq. 9.21 it is possible to obtain the value of the ratio K/η for this system, taking into account that $q_{\|\mathrm{high}}$ correspond to the minimum wavelength of the oscillation modes parallel to the columns, given by:

$$q_{\|\mathrm{high}} = 2\pi/d \qquad (9.23)$$

where d is typically the distance between consecutive molecules in a column.

Using Eqs. 9.21 and 9.23, considering the distance $d \cong 3.0$ nm, determined for the columnar phase of the octopodes (Karahaliou et al., 2008), and the cutoff frequency given by Van-Quynh et al. (2010) we obtain for K/η values of the order of $10^{-9}\mathrm{m}^2\mathrm{s}^{-1}$. This result is three to four orders of magnitude higher than those found for columnar phases of low-molecular-weight liquid crystals of discotic or biforked molecules (Cruz et al., 1996, 1998). The comparatively much higher value of K/η means that the columns, in the case of octopodes, are more rigid (with respect to bending deformations) in comparison with the columnar phases of small discotic or biforked molecules.

With respect to rotational reorientation movements of the mesogenic units described by the BPP mechanism (see Eq. 9.9), it is interesting to verify the characteristic time τ_1 follows an Arrhenius law independently of the phase. This conclusion is drawn by plotting τ_1 versus the inverse temperature for the whole range

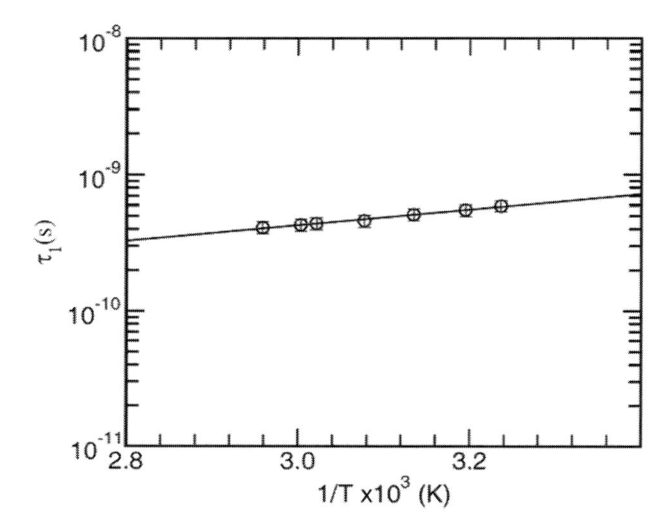

Figure 9.13 Inverse temperature dependence of τ_1 in organosiloxane octopode (nematic and Col_h phases) and fit with Arrhenius law (Eq. 9.10): $\tau_\infty \approx 8.4 \times 10^{-12}$ s, activation energy $E_a = 11$ kJ/mol. From Van-Quynh, A., Sebastiao, P. J., Wilson, D. A. and Mehl, G. H. (2010). Detecting columnar deformations in a supermesogenic octapode by proton NMR relaxometry, *European Physical Journal E* **31**, 3, pp. 275–283. With permission of Springer.

under investigation (including isotropic, nematic and Col_h phases) and fitting the data with Eq. 9.10. The result, presented in Fig. 9.13, shows that this mechanism is compatible with a thermally activated process with a extrapolated time for infinite temperature $\tau_\infty \approx 8.4 \times 10^{-12}$ s and an activation energy $E_a = 11$ kJ/mol. This value is close to the one found for Ts and Tas (see Fig. 9.9) in agreement with the assumption that the mechanism described by the BPP model correspond to the rotational reorientations of the mesogenic units, which are similar in Tas and in the octopode studied in (Van-Quynh et al., 2010).

9.1.4 PAMAM Liquid-Crystalline Dendrimers of Generations 1 and 3

To explore the relation between complexity of dendritic architectures and molecular dynamics, relaxometry studies on polyami-

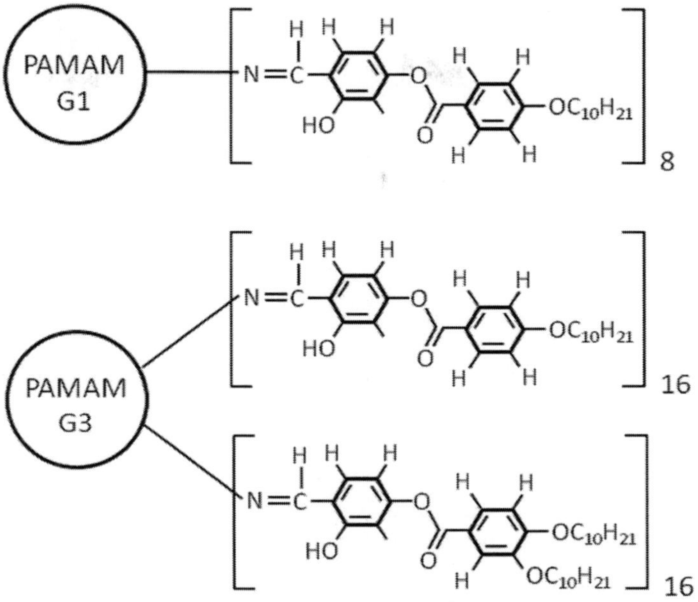

$$G = 1: \quad g - 80\ ^\circ\text{C} \rightarrow \text{SmA} \rightarrow 139\ ^\circ\text{C} \rightarrow \text{I}$$
$$G = 3: \quad g - 35.5\ ^\circ\text{C} \rightarrow \text{Col}_r \rightarrow 108.2\ ^\circ\text{C} \rightarrow \text{SmA} \rightarrow 171\ ^\circ\text{C} \rightarrow \text{I}$$

Figure 9.14 Chemical structure and phase sequences of LC PAMAM generations 1 (a) and 3 (b) (Van-Quynh et al., 2006).

doamine (PAMAM) LC dendrimers of generations 1 and 3 were carried out by Van-Quynh et al. (Van-Quynh et al., 2005, 2006). The compounds used in those studies were synthesized by the group of Jose Luis Serrano at the University of Zaragoza and their phases structural properties were investigated by Guillon, Donnio, and co-workers at IPCMS Strasbourg, these works are described in (Donnio et al., 2002) and (Rueff et al., 2003). The chemical structure of the investigate materials are presented in Fig. 9.14.

The G1 dendrimer exhibits a single smectic A phase whereas G3 LC codendrimer shows a SmA phase at higher temperature and a rectangular columnar phase for a lower temperature range. The proposed molecular packing models for the mesophases of referred

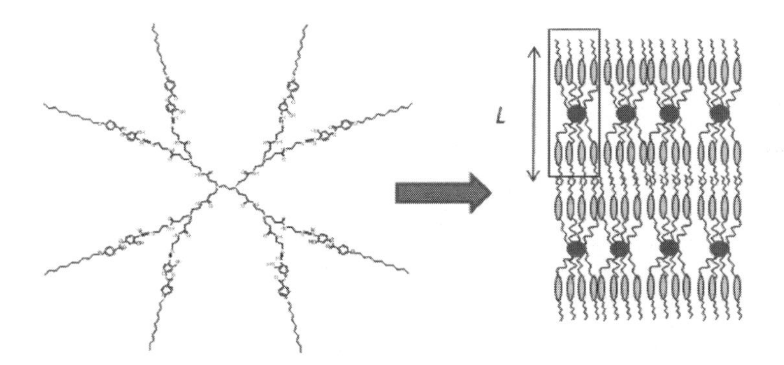

Figure 9.15 Molecular packing model of LC PAMAM generation 1. The dimension L is 5.4 nm at 110°C.

LC dendrimer compounds are schematically shown in Figs. 9.15 and 9.16 (Donnio et al., 2002) and (Rueff et al., 2003). In the case of the generation 1 LC dendrimer, in the SmA phase, the dendrimer assumes an elongated conformation (approximately cylindrical). This molecular overall shape is determined by the separation of the eighth terminal mesogenic units in two subgroups of four (in average) with the dendrimer flexible core in the middle. According to the phase structural model, determined by XRD, the SmA phase is composed of a sequence of sublayers of dendrimer cores and head-to-head mesogenic units, with aromatic rigid segments and aliphatic chains, respectively. This molecular packing model is consistent with the microsegregation effect between the PAMAM dendritic flexible core, the aromatic rigid segments, and the aliphatic chains, as schematically depicted in Fig. 9.15.

The generation 3 compound is a PAMAM codendrimer with 16 terminal mesogenic units composed of a rigid aromatic core ended by a single aliphatic chain and another 16 formed by a similar rigid aromatic core bearing two terminal aliphatic chains. As discussed in Chapter 3, the functionalization of PAMAM dendrimers with single chain mesogenic units favors the formation of elongated molecular overall shapes (for generations below the "starburst" dendrimer limit), leading in general to smectic phases. On the other hand, the end-on functionalization of the dendritic core with mesogenic units terminated with two or more aliphatic chains favors

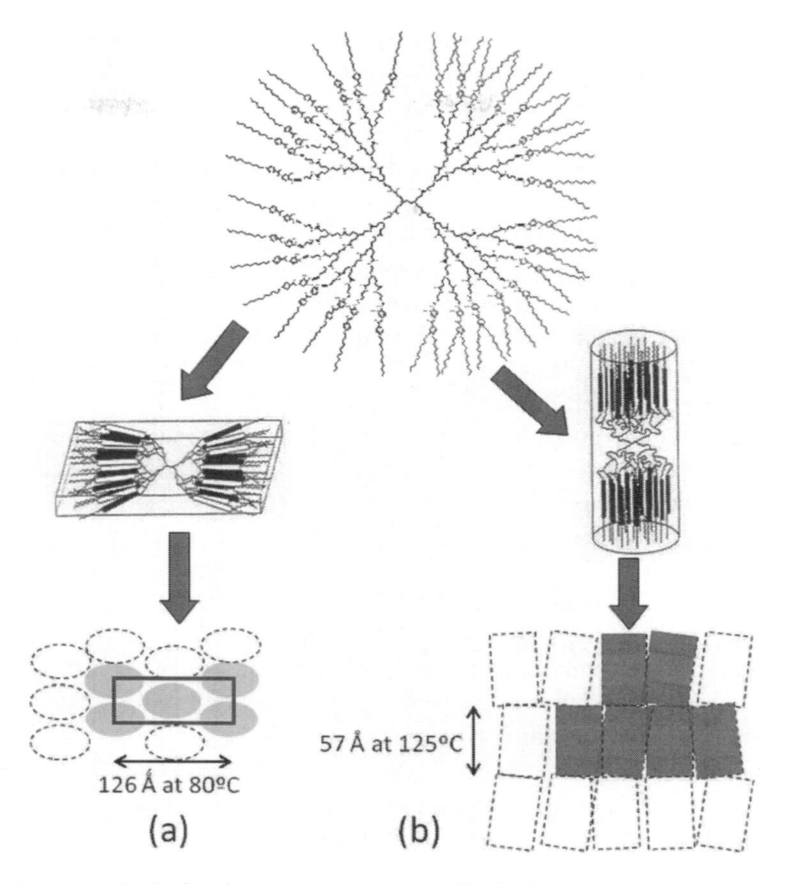

Figure 9.16 Molecular packing models of LC PAMAM codendrimer of generation 3 in (a) Col_r and (b) SmA phases.

the assumption of a disk-like shape of the dendrimers giving rise to columnar phases at suitable temperatures. The codendrimer system investigated by NMR relaxometry as reported in (Van-Quynh et al., 2005, 2006) presents the interesting feature of allowing both elongated or planar configurations of the whole dendrimer, depending on the temperature, leading to smectic or columnar molecular arrangements, respectively.

The NMR relaxometry results obtained for the PAMAM LC dendrimers described above allow for two kinds of analysis. Firstly we will compare the results obtained in the smectic A phases of G1 and G3 dendrimers respectively. This study aims to contribute to the understanding of the effect of the dendrimer generation on the molecular dynamics. It's important to notice that we are comparing phases with similar molecular organization (actually the molecular packing models are significantly alike) comprised of molecular entities of different sizes and level of complexity. Basically, comparing with the G1 LC dendrimer, the G3 LC PAMAM has a dendritic core with two additional branching levels accommodated between mesogenic sublayers that are relatively similar to those of the G1 dendrimer. The impact of these differences and similarities on the molecular dynamics behavior of such systems has been reported by Van Quynh et al. (Van-Quynh et al., 2006).

In another investigation (Van-Quynh et al., 2005) the molecular dynamics in the smectic A and columnar rectangular phases of the G3 PAMAM LC dendrimer will be compared. In that case, different spatial molecular arrangements of similar molecular entities (the G3 codendrimers) is compared. The objective of the study is to investigate the effect of the molecular configurations, the corresponding molecular packing, and the consequent short- and long-range correlations between mesogenic segments and dendritic cores on the molecular dynamics.

In Fig. 9.17 we present the T_1 results for temperatures corresponding to the SmA phases of LC PAMAM of generations 1 (a) and 3 (b), respectively. The dashed curves correspond to the theoretical relaxation models presented in Chapter 6, associated with different molecular motions. The solid curve results from the fit of the sum of the partial contributions to the experimental data according to Eq. 9.1.

The T_1 results obtained in the smectic A phase of the G1 PAMAM dendrimer are well described by the sum of three contributions for the overall relaxation rate:

$$\frac{1}{T_1} = \left(\frac{1}{T_1}\right)_{\text{BPP}} + \left(\frac{1}{T_1}\right)_{LU} \tag{9.24}$$

For high frequencies ($\omega/2\pi > 10 - 100$ MHz) the relaxation is dominated by a mechanism (BPP) described by a sum of two BPP functions, which are assigned, as usual, to rotational reorientation movements of molecules or molecular segments.

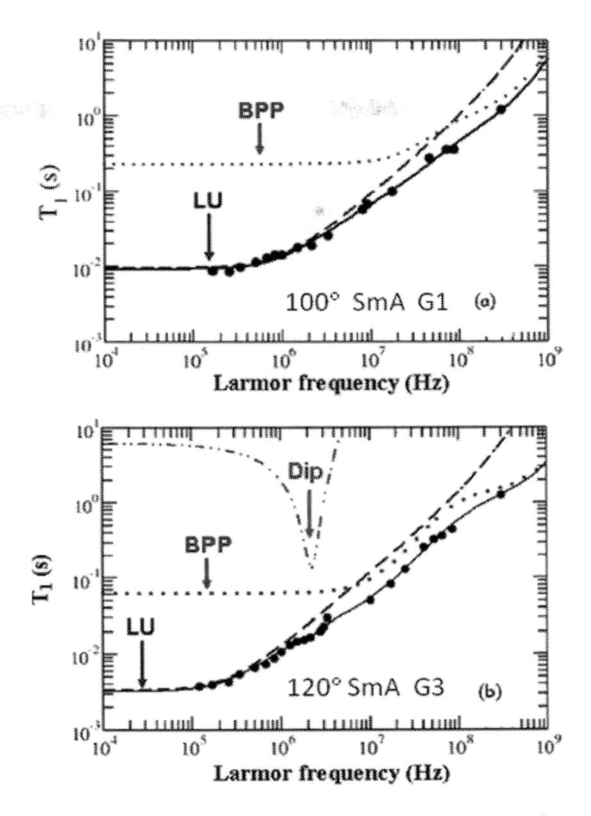

Figure 9.17 Relaxation time (T_1) as a function of the Larmor frequency and fits of relaxation models for temperatures corresponding to the SmA phase of LC PAMAM of generations G1 (100°C) and G3 (120°C). From Van-Quynh, A., Filip, D., Cruz, C., Sebastiao, P. J., Ribeiro, A. C., Rueff, J. M., Marcos, M. and Serrano, J. L. (2006). NMR relaxation study of molecular dynamics in the smectic a phase of PAMAM liquid crystalline dendrimers of generation 1 and 3, *Molecular Crystals and Liquid Crystals* **450**, pp. 391–401, reprinted by permission of Taylor & Francis Ltd, www.tandfonline.com.

These functions (see Eq. 9.9) are characterized by the factors A_s and A_f (denoting respectively slow and fast rotations) and the corresponding correlation times τ_s and τ_f.

$$\left(\frac{1}{T_1}\right)_{\text{BPP}}(\omega) = \left(\frac{1}{T_1}\right)_{\text{BPP}}(A_s, \tau_s, \omega) + \left(\frac{1}{T_1}\right)_{\text{BPP}}(A_f, \tau_f, \omega) \quad (9.25)$$

For low Larmor frequencies (100 kHz–10 MHz), the dominant mechanism is clearly identified as LUs, described by a contribution proportional to $1/\omega$ (see Eq. 9.11).

In the case of the G3 dendrimer, an additional contribution described as cross-relaxation (CR), is introduced to explain the *dip* observed in the frequency dispersion of the T_1 data:

$$\frac{1}{T_1} = \left(\frac{1}{T_1}\right)_{\text{BPP}} + \left(\frac{1}{T_1}\right)_{LU} + \left(\frac{1}{T_1}\right)_{CR} \tag{9.26}$$

This *dip* is generally explained as the effect of *cross-relaxation* between the ^1H and ^{14}N spins' populations as described in Chapter 6. In the case of the G1 dendrimer, this effect is not evident, probably due to the more reduced number of nitrogen atoms (in comparison with the ^1H population in the system) and, therefore, only the BPP and LU terms are necessary to explain the experimental data.

The cross-relaxation term is given by (see 6) (Pusiol et al., 1992):

$$\left(\frac{1}{T_1}\right)_{CR}(\omega) = D\frac{\tau_{CR}}{1 + (\omega_{CR} - \omega)^2\tau_{CR}^2} \tag{9.27}$$

where D is an amplitude factor, ω_{CR} is a resonant frequency characteristic of the so-called *quadrupole dip*, and τ_{CR} is a correlation time related to the half-height width of the Lorentzian curve described by Eq. 9.27.

With respect to the high-frequency regime (ω > 1–10 MHz), described by the sum of the two BPP contributions (see Eq. 9.25), the conclusions of the fits of Eqs. 9.24 and 9.26 lead to values of τ_f of the order of 10^{-10}s and τ_s of the order of 10^{-9}s. Those values are similar to correlation times found on low-molecular-weight liquid crystals of biforked molecules for reorientations/rotations movements around the long and short molecular axes respectively (Cruz et al., 1996). Actually, the conditions found in dendrimers are different from those of the local rotations/reorientations of low-molecular-weight liquid crystal molecules since, in dendrimers, the mesogenic units are covalently linked to the central dendritic core. Nevertheless a reasonable analogy can be established if we consider the molecular structure of the terminal units of the end-on side-chain PAMAM LC dendrimers, comparing with typical low-molecular-weight thermotropic LCs.

With respect to the low-Larmor-frequency regime, corresponding to the LU mechanism in the SmA phase of the G1 and G3 dendrimers, the most interesting result reported in (Van-Quynh et al., 2006) is obtained by comparing the low cutoff frequency ω_{cm} associated to this mechanism as a function of the dendrimer generation (see Eq. 9.11). Those values are $\omega_{cm}/(2\pi) \cong 1.3 \times 10^6$ Hz for generation 1 and $\omega_{cm}/(2\pi) \cong 2.2 \times 10^5$ Hz for generation 3, and are two and one order of magnitude greater than those of low-molecular-weight LCs (Cruz et al., 1996; Sebastião et al., 1993).

The low and high cutoff frequencies are given by

$$\omega_{cm} = K_1 q_{min}^2/\eta \tag{9.28}$$

$$\omega_{cM} = K_1 q_{max}^2/\eta \tag{9.29}$$

where K_1 is the splay elastic constant and η the corresponding viscosity, and

$$q_{min} = \frac{2\pi}{\xi_\perp} \tag{9.30}$$

$$q_{max} = \frac{2\pi}{d} \tag{9.31}$$

q_{min} is the lowest wavenumber associated with the LUs, and $\xi_\perp$ represents a coherence length that is given by the largest possible wavelength oscillation mode. q_{max} is the highest wavenumber of LUs and d is the shortest-possible wavelength, typically of the order of the lateral distance between molecules.

If we consider, as a reasonable assumption, that the ratio K_1/η has similar values for the dendrimers of generations 1 and 3, it is possible to obtain the following relation between the coherence lengths associated with dendrimers of the two generations, from the values of the low cutoff frequencies ω_c.

$$\frac{(\xi_\perp)_{G3}}{(\xi_\perp)_{G1}} = \sqrt{\frac{\omega_{c1}}{\omega_{c3}}} \cong 2.4 \tag{9.32}$$

This value is in line with the ratio between the diameter of the individual dendrimers in their ideal shape assumed in the packing model proposed for the SmA phase of the PAMAM LC dendrimers, $\phi_3/\phi_1 = 2.2$ (Barbera et al., 1999) (see Figs. 9.16 and 9.18). This result means that the maximum number of dendrimers included in the linear dimension of the highest wavelength oscillation mode (corresponding to the lowest wavenumber $q_{\min}$) is nearly equal in the SmA phases exhibited by the G3 and G1 PAMAM dendrimers.

In Fig. 9.18 we present the T_1 results for temperatures corresponding to the SmA and $\mathrm{Col_r}$ phases of LC PAMAM of generation 3, respectively. The dashed curves correspond to the theoretical relaxation models presented in Chapter 6, associated with different molecular motions. The solid curve results from the fit of the sum of the partial contributions to the experimental data according to Eq. 9.1.

Van-Quynh et al. (2005) analyze the comparison between the T_1 relaxation results in the SmA phase (discussed above) and the columnar rectangular ($\mathrm{Col_r}$) of PAMAM codendrimers of generation 3.

The theoretical expression used to fit the experimental results in the $\mathrm{Col_r}$ is similar to that corresponding to the SmA phase (see Eq. 9.26), substituting the contribution of smectic layers undulations (LU) by that of elastic columnar deformations (ECD).

$$\frac{1}{T_1} = \left(\frac{1}{T_1}\right)_{\mathrm{BPP}} + \left(\frac{1}{T_1}\right)_{ECD} + \left(\frac{1}{T_1}\right)_{CR} \tag{9.33}$$

The BPP term associated to local rotations/reorientations of the mesogenic units is described by Eq. 9.25 and, as in the SmA phase, represents the dominant contributions at high Larmor frequencies ($\omega/(2\pi) > 1$–10 MHz). Also in agreement to what is observed in the SmA phase, the values of the correlations times obtained from the fits correspond to τ_f of the order of 10^{-10} s and τ_s of the order of 10^{-9} s. The similar description and order of magnitude of these fitting parameters, both in the SmA and $\mathrm{Col_r}$ phases, is consistent with the assumption that the rotations/reorientation movements are essentially a local mechanism corresponding to the dynamics of specific molecular segments not particularly affected by the long distance properties of the mesophase organization. The values found for τ_f and τ_s, which are similar to those previously reported for low-molecular-weight liquid crystals, are an indication that the BPP contribution observed in the present case correspond to the molecular rotations/reorientation of the peripheral mesogenic units (that are similar to low-molecular-weight LCs).

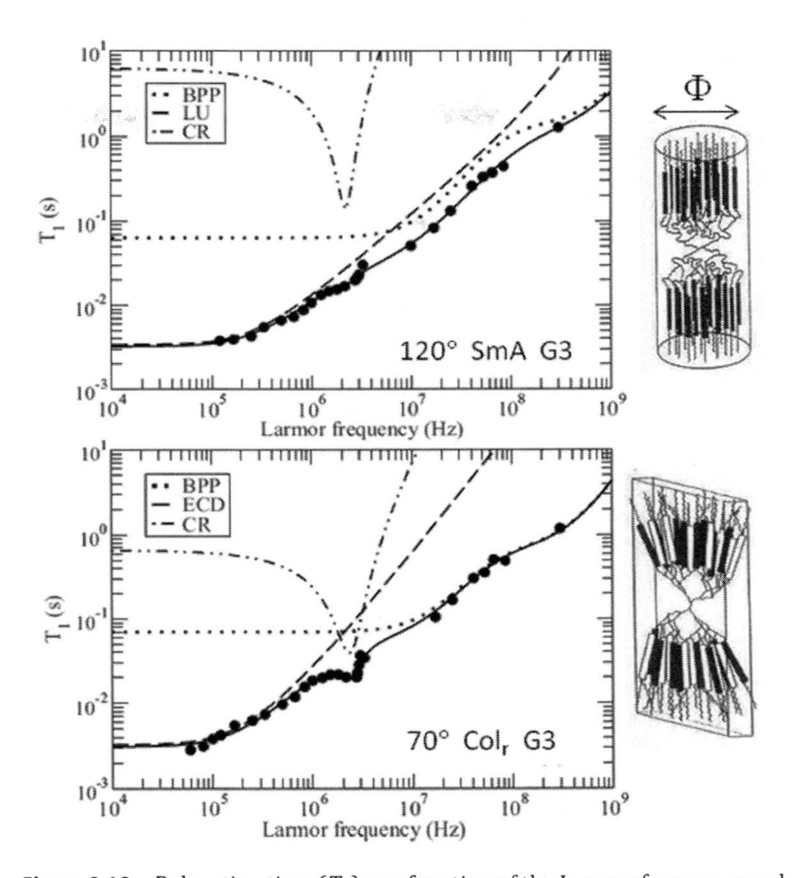

Figure 9.18 Relaxation time (T_1) as a function of the Larmor frequency and fits of relaxation models for temperatures corresponding to the SmA and Col$_r$ phases of LC PAMAM of generation G3. From Van-Quynh, A., Filip, D., Cruz, C., Sebastiao, P. J., Ribeiro, A. C., Rueff, J. M., Marcos, M. and Serrano, J. L. (2005). NMR relaxation study of molecular dynamics in columnar and smectic phases of a PAMAM liquid-crystalline codendrimer, *European Physical Journal E* **18**, 2, pp. 149–158. With permission of Springer.

The cross-relaxation term is also evident in the Col$_r$ phase, as can be observed by the *dip* in the frequency dependence of the spin–lattice relaxation time T_1 (see Fig. 9.18).

The frequency dependence in the low-Larmor-frequency domain (typically for $\omega/(2\pi) < 1$ MHz) for the Col$_r$ phase, is described by Eq. 9.20 (as previously introduced in Chapter 6). The fitting parameters are the same used in the case of the columnar phase exhibited by organosiloxane octopodes (see Eqs. 9.21, 9.22, 9.23).

As previously discussed for the Col_h of the octopodes, the viscoelastic properties of the system, that influence the collective movements, can be to some extent accessed through the analysis of the values of the high cutoff frequency ω_c obtained from the fits. In the present case (PAMAM G3) the value of $\omega_c/(2\pi)$ is close to $10^{13}s^{-1}$. This value is several orders of magnitude higher than those determined for columnar phases of discotic and biforked low-molecular-weight liquid crystals (Cruz et al., 1996, 1998) and even than that found for organosiloxane octopodes (Van-Quynh et al., 2010). Using Eqs. 9.21 and 9.23, and considering the distance between two consecutive dendrimers along the columnar axis, $d =$ 1.1nm (Van-Quynh et al., 2005) the value $K/\eta \cong 2 \times 10^{-6}m^2s^{-1}$ is obtained. This value is 3 orders of magnitude higher that than found for the Col_r phase of organosiloxane octopodes, indicating that the columnar structure formed by the PAMAM dendrimers is considerably more rigid than that exhibited by the organosiloxane octopodes.

9.2 The Influence of Dendrimers' Structure on the Molecular Dynamics

The comparison of the NMR relaxometry results on several types of LC dendrimers presented before provides a significant insight on the relation between the molecular architectures, the resulting mesophases' structural properties and the way they influence the molecular movements in a wide range of timescales (from milliseconds to picoseconds).

The most general and evident conclusion from the investigations on the side-chain dendrimers presented in this work is the absence of the contribution of molecular self-diffusion movements to the proton NMR spin–lattice relaxation rate. This is particularly interesting, considering the variety of dendrimers' molecular structures, generations and mesophases studied. The investigations include generations G0 (organosiloxane tetrapodes),

G1 (organosiloxane octopodes and PAMAM derivatives), and G3 (PAMAM LC codendrimer), and mesophases: nematics (N_u, N_b), smectics (SmA, SmA_d, SmC, SmC_d), and columnar (Col_h, Col_r). As referred before, this effect is due to the molecular interdigitation between mesogenic units belonging to neighboring dendrimers. The resulting entanglement within the dendrimers' mesophases strongly restrict diffusive movements of the whole dendrimers. Nevertheless, the diffusion of dendritic molecules, mediated by successive displacement of the dendritic arms is not excluded, while probably too slow to be experimentally observed by the method presented herein. This conclusion is particularly evident when the mesophases of the G0 dendrimers (tetrapodes) are compared with the similar phases of the corresponding monomers. It is important to remind that the relaxation results on the nematic phase of the monomers are similar to those observed in other low-molecular-weight liquid crystals, including the typical observation of the molecular self-diffusion process. However, the single (but determinant) difference between the monomers and the tetrapodes phases is the linking of the monomers to the central silicon atom (dendritic core), which determines a completely distinct NMR relaxation behavior, remarkably evident at the frequency range associated to self-diffusion. This is a clear indication that the dendritic structure is indeed responsible for the strong restriction of diffusive movements.

Also as a general conclusion, it was observed that in the high-Larmor-frequency domain (typically for $\omega/(2\pi) > 10$ MHz) the proton spin–lattice relaxation rate is dominated by contributions compatible with BPP models. This is consistent with the occurrence of molecular rotations/reorientations of the dendritic terminal mesogenic units. In particular, the order of magnitude of the characteristic correlation times and the prefactors associated to interproton distances are similar with those of low-molecular-weight liquid crystals. Moreover, in several cases, it was observed that the temperature dependences of those BPP mechanisms are compatible with Arrhenius-type laws, with values of activation energies close to those found in low molecular mass LCs. Those Arrhenius-type dependences are, in many cases, independent of the phase transition temperatures, meaning that the same Arrhenius law is valid through a temperature range including different mesophases. This is clearly an indication that the molecular rotation/reorientation mechanism is essentially defined by local molecular conditions, independent of the long distance structure of the mesophase.

Contrary to what is observed with respect to the high-frequency regime, the low-frequency dependence of the relaxation rate, which is associated with collective movements, is normally dependent of the long-distance phase organization. Typically, dependences of the type $T_1^{-1} \approx \omega^{-1/2}$ and $T_1^{-1} \approx \omega^{-1}$, are observed in nematic and smectic mesophases, respectively. In columnar phases a different type of frequency dependence, corresponding to a mechanism described as *columnar elastic deformations ECD,* is found (see Eq. 9.20).

The effect of generations was also studied by comparing the SmA phases exhibited by G1 and G3 PAMAM dendrimers. As could be expected, the contribution of local rotations/reorientations of mesogenic units is practically independent of the generation, confirming the determinant influence of the local molecular conditions on this type of movements. The collective movements are described as LUs in both 1 and 3 generation dendrimers. The low cutoff frequencies detected in the two generations indicate that the number of dendrimers contributing to the layers undulations is the roughly the same independently of the dendrimers' generation.

However, the relation between the low-frequency dependence of the relaxation rate and the long-range molecular organization is not completely strict. Actually an interesting exception becomes evident from the comparison between G0 dendrimers with terminally (end-on) and laterally attached (side-on) mesogenic units. In the case of the organosiloxane end-on tetrapodes with a strong terminal dipole referred before, partial bilayer SmA_d and SmC_d are formed and the NMR relaxation data reveal an evident contribution of the LU mechanism typically observed in low-molecular-weight liquid crystals. However, in the case of the organosiloxane tetrapodes with laterally attached mesogens, the collective movements observed in the SmC phase are similar to those detected in the nematic phase (ODF). As discussed before, this effect is also extended to the isotropic phase and is. determined by the very particular phase structure of organosiloxane tetrapodes with a peculiar temperature-persistent cybotactic structure.

Based on the NMR investigations referred in this chapter, it is possible to summarize the main conclusions on the molecular dynamics of LC dendrimers in a single sentence, by stating that molecular self-diffusion is not detected, the dendrimers molecular structure determines with particular incidence the collective movements associated with the mesophase long-range organization, and the high-frequency regime is dominated by local rotational reorientation movements of the functional mesogenic units.

Bibliography

Abragam, A. (1961). *The Principles of Nuclear Magnetism* (Oxford, University Press: Oxford, England).

Acharya, B. R., Primak, A. and Kumar, S. (2004). Biaxial nematic phase in bent-core thermotropic mesogens, *Physical Review Letters* **92**, 14, p. 145506, doi:10.1103/PhysRevLett.92.145506.

Agina, E. V., Ponomarenko, S. A., Boiko, N. I., Rebrov, E. A., Muzafarov, A. M. and Shibaev, V. P. (2001). Synthesis and phase behavior of carbosilane lc dendrimers with terminal mesogenic groups based on anisic acid derivatives, *Polymer Science Series A* **43**, 10, pp. 1000–1007.

Astruc, D., Boisselier, E. and Ornelas, C. (2010). Dendrimers designed for functions: from physical, photophysical, and supramolecular properties to applications in sensing, catalysis, molecular electronics, photonics, and nanomedicine, *Chemical Reviews* **110**, 4, pp. 1857–1959, doi:10.1021/cr900327d.

Baars, M. W. P. L., Sontjens, S. H. M., Fischer, H. M., Peerlings, H. W. I. and Meijer, E. W. (1998). Liquid-crystalline properties of poly(propylene imine) dendrimers functionalized with cyanobiphenyl mesogens at the periphery, *Chemistry - A European Journal* **4**, 12, pp. 2456–2466, doi:10.1002/(SICI)1521-3765(19981204)4:12<2456.

Barbera, J., Marcos, M. and Serrano, J. L. (1999). Dendromesogens: liquid crystal organizations versus starburst structures, *Chemistry - A European Journal* **5**, 6, pp. 1834–1840, doi:10.1002/ (SICI)1521-3765(19990604)5:6¡1834.

Barbera, J., Donnio, B., Gimenez, R., Guilllon, D., Marcos, M., Omenat, A. and Serrano, J. L. (2001). Molecular morphology and mesomorphism in dendrimers: a competition between rods and discs, *Journal of Materials Chemistry* **11**, 11, pp. 2808–2813, doi:10.1039/b102711p.

Barbera, J., Donnio, B., Gehringer, L., Guillon, D., Marcos, M., Omenat, A. and Serrano, J. L. (2005). Self-organization of nanostructured functional dendrimers, *Journal of Materials Chemistry* **15**, 38, pp. 4093–4105, doi: 10.1039/b502464a.

Beckmann, P. A., Emsley, J. W., Luckhurst, G. R. and Turner, D. L. (1983). Molecular-dynamics in a liquid-crystal: measurement of spectral densities at several sites in the nematogen 4-n-pentyl-4'-cyanobiphenyl, *Molecular Physics* **50**, 4, pp. 699–725, doi:10.1080/00268978300102631.

Beckmann, P. A., Emsley, J. W., Luckhurst, G. R. and Turner, D. L. (1986). Nuclear-spin lattice-relaxation rates in liquid-crystals: results for deuterons in specifically deuteriated 4-n-pentyl-4'-cyanobiphenyl in both nematic and isotropic phases, *Molecular Physics* **59**, 1, pp. 97–125, doi:10.1080/00268978600101941.

Blinc, R., Burgar, M., Luzar, M., Pirš, J., Zupančič I. and Žumer, S. (1974). Anisotropy of self-diffusion in smectic-A and smectic-C phases, *Physical Review Letters* **33**, 20, pp. 1192–1195, doi:10.1103/PhysRevLett.33.1192.

Blinc, R., Vilfan, M. and Rutar, V. (1975). Nature of spin-lattice relaxation in nematic MBBA, *Solid State Communications* **17**, 2, pp. 171–174, doi:10.1016/0038-1098(75)90035-6.

Blinov, L. M. and Chigrinov, V. G. (1994). *Electrooptical Effects in Liquid Crystal Materials* (Springer).

Bloembergen, N., Purcell, E. M. and Pound, R. V. (1948). Relaxation effects in nuclear magnetic resonance absorption, *Physical Review* **73**, 7, pp. 679–712, doi:10.1103/PhysRev.73.679.

Bloch, F. (1946). Nuclear induction, *Physical Review* **70**, p. 460.

Bobrovsky, A. Y., Pakhomov, A. A., Zhu, X. M., Boiko, N. I. and Shibaev, V. P. (2001). Photooptical behavior of a liquid-crystalline dendrimer of the first generation with azobenzene terminal groups, *Polymer Science Series A* **43**, 4, pp. 431–437.

Bobrovsky, A. Y., Pakhomov, A. A., Zhu, X. M., Boiko, N. I., Shibaev, V. P. and Stumpe, J. (2002). Photochemical and photoorientational behavior of liquid crystalline carbosilane dendrimer with azobenzene terminal groups, *Journal of Physical Chemistry B* **106**, 3, pp. 540–546, doi:10.1021/jp0125247.

Boiko, N., Zhu, X. M., Bobrovsky, A. and Shibaev, V. (2001). First photosensitive liquid crystalline dendrimer: synthesis, phase behavior, and photochemical properties, *Chemistry of Materials* **13**, 5, pp. 1447–1452, doi:10.1021/cm001116x.

Boiko, N. I., Lysachkov, A. I., Ponomarenko, S. A., Shibaev, V. P. and Richardson, R. M. (2005). Synthesis and comparative studies of carbosilane liquid crystalline dendrimers with chiral terminal mesogenic groups,

Colloid and Polymer Science **283**, 11, pp. 1155–1162, doi:10.1007/s00396-005-1313-6.

Bonsignore, S., Cometti, G., Dalcanale, E. and du Vosel, A. (1990). New columnar liquid-crystals correlation between molecular-structure and mesomorphic behavior, *Liquid Crystals* **8**, 5, pp. 639–649, doi:10.1080/02678299008047377.

Bos, P. J., Shetty, A., Doane, J.W. and Neubert, M.E. (1980). Molecular and relative segmental order in the nematic and smectic c-phases: a nuclear magnetic-resonance study, *Journal of Chemical Physics* **73**, pp. 733–743.

Cameron, J. H., Facher, A., Lattermann, G. and Diele, S. (1997). Poly(propyleneimine) dendromesogens with hexagonal columnar mesophase, *Advanced Materials* **9**, 5, pp. 398–403, doi:10.1002/adma.19970090507.

Campidelli, S., Eng, C., Saez, I. M., Goodby, J. W. and Deschenaux, R. (2003). Functional polypedes: chiral nematic fullerenes, *Chemical Communications* **13**, pp. 1520–1521, doi:10.1039/b303798n.

Cardoso, M., Figueirinhas, J. L., Cruz, C., Van-Quynh, A., Ribeiro, A. C., Feio, G., Apreutesei, D. and Mehl, G. H. (2008). Deuterium NMR investigation of the influence of molecular structure on the biaxial ordering of organosiloxane tetrapodes nematic phase, *Molecular Crystals and Liquid Crystals* **495**, pp. 348–359, doi:10.1080/15421400802430489.

Carvalho, A., Sebastião, P. J., Ribeiro, A. C., Nguyen, H. T. and Vilfan, M. (2001a). Molecular dynamics in tilted bilayer smectic phases: a proton nuclear magnetic resonance relaxometry study, *Journal of Chemical Physics* **115**, 22, pp. 10484–10492.

Carvalho, A., Sebastião, P. J., Ribeiro, A. C., Nguyen, H. T. and Vilfan, M. (2001b). Molecular dynamics in tilted bilayer smectic phases: a proton nuclear magnetic resonance relaxometry study, *Journal of Chemical Physics* **115**, 22, pp. 10484–10492.

Chandrasekhar, S. and Ranganath, G. S. (1990). Discotic liquid crystals, *Reports on Progress in Physics* **53**, 1, p. 57, http://stacks.iop.org/0034-4885/53/i=1/a=002.

Channabasaveshwar, V., Prasad, Y. S. K., Nair, G. G., Shashikala, I. S., Rao, D. S. S., Lobo, C. V. and Chandrasekhar, S. (2004). A low-molar-mass, monodispersive, bent-rod dimer exhibiting biaxial nematic and smectic a phases, *Angewandte Chemie-international Edition* **43**, 26, pp. 3429–3432, doi:10.1002/anie.200453908.

Chuard, T., Beguin, M. T. and Deschenaux, R. (2003). Ferrocene-containing liquid-crystalline dendrimers, *Comptes Rendus Chimie* **6**, 8-10, pp. 959–962, doi:10.1016/j.crci.2003.05.002.

Coen, M. C., Lorenz, K., Kressler, J., Frey, H. and Mulhaupt, R. (1996). Mono- and multilayers of mesogen-substituted carbosilane dendrimers on mica, *Macromolecules* **29**, 25, pp. 8069–8076, doi:10.1021/ma951545u.

Cohen-Tannoudji, C., Diu, B. and Laloe, F. (1977). *Quantum Mechanics* (John Wiley & Sons, Hermann).

Cook, A. G., Baumeister, U. and Tschierske, C. (2005). Supramolecular dendrimers: unusual mesophases of ionic liquid crystals derived from protonation of dab dendrimers with facial amphiphilic carboxylic acids, *Journal of Materials Chemistry* **15**, 17, pp. 1708–1721, doi:10.1039/b415892j.

Cordoyiannis, G., Apreutesei, D., Mehl, G. H., Glorieux, C. and Thoen, J. (2008). High-resolution calorimetric study of a liquid crystalline organo-siloxane tetrapode with a biaxial nematic phase, *Physical Review E* **78**, 1, p. 011708, doi:10.1103/PhysRevE.78.011708.

Crampton, H. L. and Simanek, E. E. (2007). Dendrimers as drug delivery vehicles: non-covalent interactions of bioactive compounds with dendrimers, *Polymer International* **56**, 4, pp. 489–496, doi:10.1002/pi.2230.

Crawford, G. P. and Woltman, S. J. (2007). *Liquid Crystals Frontiers in Biomedical Applications* (World Scientific Co. Pte. Ltd., Singapore).

Cruz, C., Figueirinhas, J. L., Filip, D., Feio, G., Ribeiro, A. C., Frere, Y., Meyer, T. and Mehl, G. H. (2008). Biaxial nematic order and phase behavior studies in an organosiloxane tetrapode using complementary deuterium NMR experiments, *Physical Review E* **78**, 5, p. 051702, doi:10.1103/PhysRevE.78.051702.

Cruz, C., Figueirinhas, J. L., Sebastião, P. J., Ribeiro, A. C., Noack, F., Nguyen, H. T., Heinrich, B. and Guillon, D. (1996). Molecular dynamics in the columnar and lamellar mesophases of a liquid crystal of biforked molecules, *Zeitschrift Fur Naturforschung Section A-A Journal Of Physical Sciences* **51**, 3, pp. 155–166.

Cruz, C., Sebastião, P. J., Figueirinhas, J. L., Ribeiro, A. C., Nguyen, H. T., Destrade, C. and Noack, F. (1998). NMR study of molecular dynamics in a dho columnar mesophase, *Z. Naturforsch.* **53a**, pp. 823–827.

Dardel, B., Deschenaux, R., Even, M. and Serrano, E. (1999). Synthesis, characterization, and mesomorphic properties of a mixed [60]fullerene-ferrocene liquid-crystalline dendrimer, *Macromolecules* **32**, 16, pp. 5193–5198, doi:10.1021/ma990034v.

de Gennes, P. and Prost, J. (1993). *The Physics of Liquid Crystals* (Oxford, University Press: Oxford, England).

Deschenaux, R., Serrano, E. and Levelut, A. M. (1997). Ferrocene-containing liquid-crystalline dendrimers: a novel family of mesomorphic macromolecules, *Chemical Communications* , 16, pp. 1577–1578, doi:10.1039/a702850d.

Deschenaux, R., Donnio, B. and Guillon, D. (2007). Liquid-crystalline fullerodendrimers, *New Journal of Chemistry* **31**, 7, pp. 1064–1073, doi:10.1039/b617671m.

Dingemans, T. J., Madsen, L. A., Francescangeli, O., Vita, F., Photinos, D. J., Poon, C.-D. and Samulski, E. T. (2013). The biaxial nematic phase of oxadiazole biphenol mesogens, *Liquid Crystals* **40**, 12, pp. 1655–1677, doi:10.1080/02678292.2013.824119.

Doane, J. W., Tarr, C. E. and Nickerson, M. (1974). Nuclear-spin-lattice relaxation in liquid-crystals by fluctuations in nematic director, *Physical Review Letters* **33**, 11, pp. 620–624, doi:10.1103/PhysRevLett.33.620.

Dolinšek, J., Jarh, O., Vilfan, M., Žumer, S., Blinc, R., Doane, J. W. and Crawford, G. (1991). 2-dimensional deuteron nuclear-magnetic-resonance of a polymer dispersed nematic liquid-crystal, *Journal of Chemical Physics* **95**, 3, pp. 2154–2161.

Dong, R. (1997). *Nuclear Magnetic Resonance of Liquid Crystals*, 2nd edn. (Springer-Verlag, New York).

Donnio, B. (2002). Lyotropic metallomesogens, *Current Opinion In Colloid & Interface Science* **7**, 5-6, pp. 371–394, doi:10.1016/S1359-0294(02)00084-5.

Donnio, B., Barbera, J., Gimenez, R., Guillon, D., Marcos, M. and Serrano, J. L. (2002). Controlled molecular conformation and morphology in poly(amidoamine) (PAMAM) and poly(propyleneimine) (DAB) dendrimers, *Macromolecules* **35**, 2, pp. 370–381, doi:10.1021/ma010881+.

Donnio, B., Buathong, S., Bury, I. and Guillon, D. (2007). Liquid crystalline dendrimers, *Chemical Society Reviews* **36**, 9, pp. 1495–1513, doi:10.1039/b605531c.

Donnio, B. and Guillon, D. (2006). Liquid crystalline dendrimers and polypedes, in *Supramolecular Polymers Polymeric Betains Oligomers,* **201** (Springer), pp. 45–155, doi:10.1007/12_079.

Donnio, B., Guillon, D., Deschenaux, R. and Bruce, D. W. (2003). *Comprehensive Coordination Chemistry II*, Vol. 7, chap. 7.9 (Elsevier, Oxford, UK), pp. 357–627.

Eidenschink, R., Kreuzer, F. H. and de Jeu, W. H. (1990). Liquid-crystalline behavior of molecules with tetrahedral symmetry, *Liquid Crystals* **8**, 6, pp. 879–884, doi:10.1080/02678299008047398.

Elsasser, R., Mehl, G. H., Goodby, J. W. and Veith, M. (2001). Nematic dendrimers based on carbosilazane cores, *Angewandte Chemie-international Edition* **40**, 14, pp. 2688–2690, doi:10.1002/1521-3773(20010716)40:14<2688.

Elsasser, R., Goodby, J. W., Mehl, G. H., Rodriguez-Martin, D., Richardson, R. M., Photinos, D. J. and Veith, M. (2003). Structure-properties relationships in a series of liquid crystals based on carbosilazane cores, *Molecular Crystals and Liquid Crystals* **402**, pp. 237–243, doi:10.1080/744816667.

Emsley, J. W. (1983). *Nuclear Magnetic Resonance of Liquid Crystals* (D. Reidel).

Ernst, R. (1976). 2-dimensional spectroscopy-application to nuclear magnetic-resonance, *Journal of Chemical Physics* **64**, pp. 2229–2246.

Ernst, R. and Anderson, W. (1966). Application of fourier transform spectroscopy to magnetic resonance, *Rev iew of Scientific Instruments* **37**, p. 93.

Ernst, R. R., Bodenhausen, G. and Wokaun, A. (1992). *Principles of Nuclear Magnetic Resonance in One and Two Dimensions* (Oxford Science).

Fabbri, U. and Zannoni, C. (1986). A monte-carlo investigation of the lebwohl-lasher lattice model in the vicinity of its orientational phase-transition, *Molecular Physics* **58**, 4, pp. 763–788, doi:10.1080/00268978600101561.

Farrar, T. C. and Becker, E. D. (1971). *Pulse and Fourrier Transform NMR* (Academic Press, New York, London).

Felekis, T., Tsiourvas, D., Tziveleka, L. and Paleos, C. M. (2005). Hydrogen-bonded liquid crystals derived from supramolecular complexes of pyridylated poly(propyleneimine) dendrimers and a cholesterol-based carboxylic acid, *Liquid Crystals* **32**, 1, pp. 39–43, doi:10.1080/02678290412331320566.

Figueirinhas, J. L., Cruz, C., Filip, D., Feio, G., Ribeiro, A. C., Frere, Y., Meyer, T. and Mehl, G. H. (2005). Deuterium NMR investigation of the biaxial nematic phase in an organosiloxane tetrapode, *Physical Review Letters* **94**, 10, p. 107802, doi:10.1103/PhysRevLett.94.107802.

Figueirinhas, J. L., Cruz, C., Feio, G. and Mehl, G. H. (2009). Collective modes and biaxial ordering observed by deuterium NMR in the nematic phases of an organosiloxane tetrapode, *Molecular Crystals and Liquid Crystals* **510**, pp. 158–174, doi:10.1080/15421400903060888.

Filip, D., Cruz, C., Sebastião, P. J., Ribeiro, A. C., Meyer, T. and Mehl, G. H. (2005). Peculiar molecular dynamics behaviour in the isotropic phase of some liquid crystalline systems, *Molecular Crystals and Liquid Crystals* **436**, pp. 17/[971]28/[982].

Filip, D., Cruz, C., Sebastião, P. J., Ribeiro, A. C., Vilfan, M., Meyer, T., Kouwer, P. H. J. and Mehl, G. H. (2007). Structure and molecular dynamics of the mesophases exhibited by an organosiloxane tetrapode with strong polar terminal groups, *Physical Review E* **75**, 1, p. 011704, doi:10.1103/PhysRevE.75.011704.

Filip, D., Cruz, C., Sebastião, P. J., Cardoso, M., Ribeiro, A. C., Vilfan, M., Meyer, T., Kouwer, P. H. J. and Mehl, G. H. (2010). Phase structure and molecular dynamics of liquid-crystalline side-on organosiloxane tetrapodes, *Physical Review E* **81**, 1, p. 011702, doi:10.1103/PhysRevE.81.011702.

Francescangeli, O. and Samulski, E. T. (2010). Insights into the cybotactic nematic phase of bent-core molecules, *Soft Matter* **6**, 11, pp. 2413–2420, doi:10.1039/c003310c.

Freed, J. H. (1977). Stochastic-molecular theory of spin-relaxation for liquid-crystals, *Journal of Chemical Physics* **66**, 9, pp. 4183–4199, doi:10.1063/1.434495.

Freiser, M. J. (1970). Ordered states of a nematic liquid, *Physical Review Letters* **24**, 19, pp. 1041–1043, doi:10.1103/PhysRevLett.24.1041.

Galerne, Y. (1988). Biaxial nematics, *Molecular Crystals and Liquid Crystals* **165**, pp. 131–149, doi:10.1080/00268948808082199.

Gehringer, L., Guillon, D. and Donnio, B. (2003). Liquid crystalline octopus: an alternative class of mesomorphic dendrimers, *Macromolecules* **36**, 15, pp. 5593–5601, doi:10.1021/ma034038i.

Gehringer, L., Bourgogne, C., Guillon, D. and Donnio, B. (2004). Liquid-crystalline octopus dendrimers: block molecules with unusual mesophase morphologies, *Journal of the American Chemical Society* **126**, 12, pp. 3856–3867, doi:10.1021/ja031506v.

Gonçalves, L. N. (2004). FEATPOST and a review of 3d METAPOST packages, in *Tex, Xml, and Digital Typography,* Vol. 3130 of the series *Lecture Notes in Computer Science*, pp. 112–124, doi:10.1007/978-3-540-27773-6_8.

Gehringer, L., Bourgogne, C., Guillon, D. and Donnio, B. (2005). Main-chain liquid-crystalline dendrimers based on amido-core moieties: effect of the core structure, *Journal of Materials Chemistry* **15**, 17, pp. 1696–1703, doi:10.1039/b416953k.

Goodby, J. W., Gray, G. W., Demus, D., Goodby, J., Gray, G. W., Spiess, H.-W. and Vill, V. (2008). *Guide to the Nomenclature and Classification of Liquid Crystals* (Wiley-VCH Verlag GmbH), pp. 17–23, doi:10.1002/9783527619276.ch2a, http://dx.doi.org/10.1002/9783527619276.ch2a.

Gorman, C. (1998). Metallodendrimers: structural diversity and functional behavior, *Advanced Materials* **10**, 4, pp. 295–309, doi:10.1002/(SICI)1521-4095(199803)10:4<295.

Gradišek, A., Sebastião, P. J., Fernandes, S. N., Apih, T., Godinho, M. H. and Seliger, J. (2014). (1)h-(2)h cross-relaxation study in a partially deuterated nematic liquid crystal, *Journal of Physical Chemistry B* **118**, 20, pp. 5600–5607, doi:10.1021/jp502542q.

Gray, G. W., Demus, D., Goodby, J., Gray, G. W., Spiess, H.-W. and Vill, V. (2008). *Introduction and Historical Development* (Wiley-VCH Verlag GmbH), pp. 1–16, doi:10.1002/9783527619276.ch1a, http://dx.doi.org/10.1002/9783527619276.ch1a.

Guillon, D. (1999). *Columnar Order in Thermotropic Mesophases in Structure and Bonding*, Liquid Crystals II (Springer Verlag, Berlin Heidelberg).

Guillon, D., Heinrich, B., Ribeiro, A. C., Cruz, C. and Nguyen, H. T. (1998). Thermotropic lamellar-to-columnar phase transition exhibited by a biforked compound, *Molecular Crystals and Liquid Crystals* **317**, pp. 51–64, doi:10.1080/10587259808047105.

Hahn, H., Keith, C., Lang, H., Reddy, R. A. and Tschierske, C. (2006). First example of a third-generation liquid-crystalline carbosilane dendrimer with peripheral bent-core mesogenic units: understanding of "dark conglomerate phases," *Advanced Materials* **18**, 19, pp. 2629–2633, doi: 10.1002/adma.200600161.

Halle, B. (1994). Surface forces, undulating bilayers, and nuclear-spin relaxation, *Physical Review E* **50**, 4, pp. R2415–R2418.

Haristoy, D., Mery, S., Heinrich, B., Mager, L., Nicoud, J. F. and Guillon, D. (2000). Structure and photoconductive behaviour of a sanidic liquid crystal, *Liquid Crystals* **27**, 3, pp. 321–328.

Harmon, J. F. and Muller, B. H. (1969). *Physical Review* **182**, p. 400.

Hawker, C. J. and Frechet, J. M. J. (1990). Preparation of polymers with controlled molecular architecture: a new convergent approach to dendritic macromolecules, *Journal of the American Chemical Society* **112**, 21, pp. 7638–7647, doi:10.1021/ja00177a027.

Hsi, S., Zimmermann, H. and Luz, Z. (1978). Deuterium magnetic-resonance of some polymorphic liquid-crystals: conformation of aliphatic end chains, *Journal of Chemical Physics* **69**, pp. 4126–4146.

Huang, K. (1987). *Statistical Physics* (John Wiley & Sons).

IUPAC (1997). *IUPAC. Compendium of Chemical Terminology*, ("Gold Book") 2nd edn. (Blackwell Scientific Publications, Oxford), doi:10.1351/goldbook.

Karahaliou, P. K., Kouwer, P. H. J., Meyer, T., Mehl, G. H. and Photinos, D. J. (2007). Columnar phase structures of an organic-inorganic hybrid functionalized with eight calamitic mesogens, *Soft Matter* **3**, 7, pp. 857–865, doi:10.1039/b617696h.

Karahaliou, P. K., Kouwer, P. H. J., Meyer, T., Mehl, G. H. and Photinos, D. J. (2008). Long- and short-range order in the mesophases of laterally substituted calamitic mesogens and their radial octapodes, *Journal of Physical Chemistry B* **112**, 21, pp. 6550–6556, doi:10.1021/jp712182v.

Kats, E. I. (1978). A model of a liquid crystal, *Soviet Journal of Experimental and Theoretical Physics* **48**, p. 916.

Kim, Y.-K., Senyuk, B., Shin, S.-T., Kohlmeier, A., Mehl, G. H. and Lavrentovich, O. D. (2014). Surface alignment, anchoring transitions, optical properties, and topological defects in the thermotropic nematic phase of organo-siloxane tetrapodes, *Soft Matter* **10**, 3, pp. 500–509, doi: 10.1039/c3sm52249k.

Kimmich, R. (1997). *NMR Tomography, Diffusometry, Relaxometry* (Springer, Berlin).

Kimmich, R. and Anoardo, E. (2004). Field-cycling NMR relaxometry, *Progress in Nuclear Magnetic Resonance Spectroscopy* **44**, pp 257–320, doi:10.1016/j.pnmrs.2004.03.002.

Kimmich, R. and Weber, H. W. (1993). NMR relaxation and the orientational structure factor, *Physical Review B* **47**, 18, pp. 11788–11794, doi:10.1103/PhysRevB.47.11788.

Klenin, V. J., Panina, Y. V., Yarotskii, V. I., Ponomarenko, S. A., Boiko, N. I. and Shibaev, V. P. (2001). Phase diagram of the third-generation liquid crystalline dendrimer-carbon tetrachloride system, *Polymer Science Series A* **43**, 5, pp. 519–524.

Kowalewski, J. and Mäler, L. (2006). *Nuclear Spin Relaxation in Liquids: Theory, Experiments, and Applications* (Taylor and Francis).

K. Wuthrich (1983). Application of phase sensitive two-dimensional correlated spectroscopy (cosy) for measurements of h-1-h-1 spin-spin coupling-constants in proteins, *Biochemical and Biophysical Research Communications* **113**, pp. 967–974.

Landau, L. D., Pitaevskii, L. P., Kosevich, A. M. and Lifshitz, E. M. (1986). *Theory of Elasticity, Theoretical Physics*, Vol. 7, 3rd edn. (Butterworth-Heinemann).

Lehmann, M., Schartel, B., Hennecke, M. and Meier, H. (1999). Dendrimers consisting of stilbene or distyrylbenzene building blocks synthesis and stability, *Tetrahedron* **55**, 47, pp. 13377–13394, doi:10.1016/S0040-4020(99)00823-6.

Lehmann, O. (1889). Über fliessende krystalle, *Zeitschrift für Physikalische Chemie* **4**, pp. 462–472.

Lenoble, J., Campidelli, S., Maringa, N., Donnio, B., Guillon, D., Yevlampieva, N. and Deschenaux, R. (2007). Liquid-crystalline janus-type fullerodendrimers displaying tunable smectic-columnar mesomorphism, *Journal of the American Chemical Society* **129**, 32, pp. 9941–9952, doi:10.1021/ja071012o.

Lenoble, J., Maringa, N., Campidelli, S., Donnio, B., Guillon, D. and Deschenaux, R. (2006). Liquid-crystalline fullerodendrimers which display columnar phases, *Organic Letters* **8**, 9, pp. 1851–1854, doi: 10.1021/ol0603920.

Leung, K. M. and Lin, L. (1987). Phase-transitions of bowlic liquid-crystals, *Molecular Crystals and Liquid Crystals* **146**, pp. 71–76, doi:10.1080/00268948708071803.

Levelut, A. M. (1983). Structures of mesophases of disc-like molecules, *Journal de Chimie Physique et de Physico-chimie Biologique* **80**, 1, pp. 149–161.

Li, J. F., Crandall, K. A., Chu, P. W., Percec, V., Petschek, R. G. and Rosenblatt, C. (1996). Dendrimeric liquid crystals: isotropic-nematic pretransitional behavior, *Macromolecules* **29**, 24, pp. 7813–7819, doi: 10.1021/ma961116b.

Lin, L. (1987). Bowlic liquid-crystals, *Molecular Crystals and Liquid Crystals* **146**, pp. 41–54.

Lowe, I. and Norberg, R. (1957). Free-induction decays in solids, *Physical Review* **107**, p. 46.

Lubensky, T. C. (1970). Molecular description of nematic liquid crystals, *Physical Review A-general Physics* **2**, 6, pp. 2497–2514, doi:10.1103/PhysRevA.2.2497.

Luckhurst, G. R. (2001). Biaxial nematic liquid crystals: fact or fiction? *Thin Solid Films* **393**, 1-2, pp. 40–52, doi:10.1016/S0040-6090(01)01091-4.

Lydon, J. (2010). Chromonic review, *Journal of Materials Chemistry* **20**, 45, pp. 10071–10099, doi:10.1039/b926374h.

Madsen, L. A., Dingemans, T. J., Nakata, M. and Samulski, E. T. (2004). Thermotropic biaxial nematic liquid crystals, *Physical Review Letters* **92**, 14, p. 145505, doi:10.1103/PhysRevLett.92.145505.

Malthete, J. and Collet, A. (1985). Liquid-crystals with a cone-shaped cyclotriveratrylene core, *Nouveau Journal De Chimie-new Journal of Chemistry* **9**, 3, pp. 151–153.

Malthete, J., Levelut, A. M. and Tinh, N. H. (1985). Phasmids: a new class of liquid-crystals, *Journal de Physique Lettres* **46**, 18, pp. L875–L880.

Malthete, J., Tinh, N. H. and Levelut, A. M. (1986). New mesogens with 6-paraffinic, 4-paraffinic, or 3-paraffinic chains, *Journal of the Chemical Society - Chemical Communications* **20**, pp. 1548–1549, doi:10.1039/c39860001548.

Marcos, M., Gimenez, R., Serrano, J. L., Donnio, B., Heinrich, B. and Guillon, D. (2001). Dendromesogens: liquid crystal organizations of poly(amidoamine) dendrimers versus starburst structures, *Chemistry - A European Journal* **7**, 5, pp. 1006–1013, doi:10.1002/1521-3765(20010302)7:5<1006.

Marcos, M., Omenat, A. and Serrano, J. L. (2003). Structure-mesomorphism relationship in terminally functionalised liquid crystal dendrimers, *Comptes Rendus Chimie* **6**, 8-10, pp. 947–957, doi:10.1016/j.crci.2003.05.001.

Martin-Rapun, R., Marcos, M., Omenat, A., Barbera, J., Romero, P. and Serrano, J. L. (2005). Ionic thermotropic liquid crystal dendrimers, *Journal of the American Chemical Society* **127**, 20, pp. 7397–7403, doi:10.1021/ja042264h.

Martin-Rapun, R., Marcos, M., Omenat, A., Serrano, J. L., Luckhurst, G. R. and Mainal, A. (2004). Poly(propyleneimine) liquid crystal codendrimers bearing laterally and terminally attached promesogenic groups, *Chemistry of Materials* **16**, 24, pp. 4969–4979, doi:10.1021/cm049191l.

McKenna, M. D., Barbera, J., Marcos, M. and Serrano, J. L. (2005). Discotic liquid crystalline poly(propylene imine) dendrimers based on triphenylene, *Journal of the American Chemical Society* **127**, 2, pp. 619–625, doi:10.1021/ja0455906.

Mehl, G. H. and Saez, I. M. (1999). Polyhedral liquid crystal silsesquioxanes, *Applied Organometallic Chemistry* **13**, 4, pp. 261–272, doi:10.1002/(SICI)1099-0739(199904)13:4<261.

Meier, H., Lehmann, M. and Kolb, U. (2000). Stilbenoid dendrimers, *Chemistry - A European Journal* **6**, 13, pp. 2462–2469, doi:10.1002/1521-3765(20000703)6:13<2462.

Merkel, K., Kocot, A., Vij, J. K., Korlacki, R., Mehl, G. H. and Meyer, T. (2004). Thermotropic biaxial nematic phase in liquid crystalline organo-siloxane tetrapodes, *Physical Review Letters* **93**, 23, p. 237801, doi:10.1103/PhysRevLett.93.237801.

Neupane, K., Kang, S. W., Sharma, S., Carney, D., Meyer, T., Mehl, G. H., Allender, D. W., Kumar, S. and Sprunt, S. (2006). Dynamic light scattering study of biaxial ordering in a thermotropic liquid crystal, *Physical Review Letters* **97**, 20, p. 207802, doi:10.1103/PhysRevLett.97.207802.

Newkome, G. R., He, E. F. and Moorefield, C. N. (1999). Suprasupermolecules with novel properties: metallodendrimers, *Chemical Reviews* **99**, 7, pp. 1689–1746, doi:10.1021/cr9800659.

Newkome, G. R., Moorefield, C. M. and Vögtle, F. (2001). *Dendrimers and Dendrons. Concepts, Synyheses, Applications* (Wiley-VCH, Weinheim, New York, Chichester, Brisbane, Singapore, Toronto.).

Noack, F. (1986). NMR field-cycling spectroscopy: principles and applications, *Progress in Nuclear Magnetic Resonance Spectroscopy* **18**, 3, pp. 171–276.

Pastor, L., Barbera, J., McKenna, M., Marcos, M., Martin-Rapun, R., Serrano, J. L., Luckhurst, G. R. and Mainal, A. (2004). End-on and side-on nematic liquid crystal dendrimers, *Macromolecules* **37**, 25, pp. 9386–9394, doi: 10.1021/ma048450p.

Pelzl, G., Diele, S. and Weissflog, W. (1999). Banana-shaped compounds: a new field of liquid crystals, *Advanced Materials* **11**, 9, pp. 707–724, doi:10.1002/(SICI)1521-4095(199906)11:9<707.

Percec, V., Chu, P. W., Ungar, G. and Zhou, J. P. (1995). Rational design of the first nonspherical dendrimer which displays calamitic nematic and smectic thermotropic liquid-crystalline phases, *Journal of the American Chemical Society* **117**, 46, pp. 11441–11454, doi:10.1021/ja00151a008.

Percec, V., Mitchell, C. M., Cho, W. D., Uchida, S., Glodde, M., Ungar, G., Zeng, X. B., Liu, Y. S., Balagurusamy, V. S. K. and Heiney, P. A. (2004). Designing libraries of first generation ab(3) and ab(2) self-assembling dendrons via the primary structure generated from combinations of (ab)(y)-ab(3) and (ab)(y)-ab(2) building blocks, *Journal of the American Chemical Society* **126**, 19, pp. 6078–6094, doi:10.1021/ja049846j.

Percec, V., Won, B. C., Peterca, M. and Heiney, P. A. (2007). Expanding the structural diversity of self-assembling dendrons and supramolecular dendrimers via complex building blocks, *Journal of the American Chemical Society* **129**, 36, pp. 11265–11278, doi:10.1021/ja073714j.

Peroukidis, S. D., Karahaliou, P. K., Vanakaras, A. G. and Photinos, D. J. (2009). Biaxial nematics: symmetries, order domains and field-induced phase transitions, *Liquid Crystals* **36**, 6-7, pp. 727–737, doi:10.1080/02678290902814700.

Pesak, D. J. and Moore, J. S. (1997). Columnar liquid crystals from shape-persistent dendritic molecules, *Angewandte Chemie, International Edition* **36**, 15, pp. 1636–1639, doi:10.1002/anie.199716361.

Petrov, A. G. (1999). *The Lyotropic State of Matter, Molecular Physics and Living Matter Physics* (Gordon and Breach Science, Amsterdam).

Pincus, P. (1969). Nuclear relaxation in a nematic liquid crystal, *Solid State Communications* **7**, 4, pp. 415–&, doi:10.1016/0038-1098(69)90886-2.

Purcell, E. M. (1946). Resonance absorption by nuclear magnetic moments in a solid, *Physical Review* **69**, pp. 37–38.

Ponomarenko, S. A., Rebrov, E. A., Bobrovsky, A. Y., Boiko, N. I., Muzafarov, A. M. and Shibaev, V. P. (1996). Liquid crystalline carbosilane dendrimers: first generation, *Liquid Crystals* **21**, 1, pp. 1–12, doi:10.1080/02678299608033789.

Ponomarenko, S., Boiko, N., Rebrov, E., Muzafarov, A., Whitehouse, I., Richardson, R. and Shibaev, V. (1999). Synthesis, phase behaviour and structure of liquid crystalline carbosilane dendrimers with methoxyphenyl benzoate terminal mesogenic groups, *Molecular Crystals and Liquid Crystals Science and Technology. Section A. Molecular Crystals and Liquid Crystals* **332**, pp. 343–350.

Ponomarenko, S. A., Boiko, N. I., Shibaev, V. P. and Magonov, S. N. (2000a). Atomic force microscopy study of structural organization of carbosilane liquid crystalline dendrimer, *Langmuir* **16**, 12, pp. 5487–5493, doi:10.1021/la991661g.

Ponomarenko, S. A., Boiko, N. I., Shibaev, V. P., Richardson, R. M., Whitehouse, I. J., Rebrov, E. A. and Muzafarov, A. M. (2000b). Carbosilane liquid crystalline dendrimers: from molecular architecture to supramolecular nanostructures, *Macromolecules* **33**, 15, pp. 5549–5558, doi:10.1021/ma0001032.

Ponomarenko, S. A., Agina, E. V., Boiko, N. I., Rebrov, E. A., Muzafarov, A. M., Richardson, R. M. and Shibaev, V. P. (2001a). Liquid crystalline carbosilane dendrimers with terminal phenyl benzoate mesogenic groups: influence of generation number on phase behaviour, *Molecular Crystals and Liquid Crystals* **364**, pp. 93–100.

Ponomarenko, S. A., Boiko, N. I., Zhu, X. M., Agina, E. V., Shibaev, V. P. and Magonov, S. N. (2001b). Atomic force microscopy study of single macromolecules and the nanostructure of carbosilane liquid crystalline dendrimer mono- and multilayers, *Polymer Science Series A* **43**, 3, pp. 245–257.

Polineni, S., Figueirinhas, J. L., Cruz, C., Wilson, D. A. and Mehl, G. H. (2013). Capacitance and optical studies of elastic and dielectric properties in an organosiloxane tetrapode exhibiting a n-b phase, *Journal of Chemical Physics* **138**, 12, p. 124904, doi:10.1063/1.4795582.

Prost, J. (1984). The smectic state, *Advances In Physics* **33**, 1, pp. 1–46, doi: 10.1080/00018738400101631.

Pusiol, D. J., Humpfer, R. and Noack, F. (1992). Nitrogen nuclear-quadrupole resonance dips in the proton spin relaxation dispersion of nematic and smectic thermotropic liquid-crystals, *Zeitschrift Fur Naturforschung Section A - A Journal of Physical Sciences* **47**, 11, pp. 1105–1114.

Rabi, I. I. (1937). Space quantization in a gyrating magnetic field, *Physical Review* **51**, p. 652.

Ramakrishnan, V. and Moore, P. B. (2001). Atomic structures at last: the ribosome in 2000, *Current Opinion In Structural Biology* **11**, 2, pp. 144–154, doi:10.1016/S0959-440X(00)00184-6.

Ramzi, A., Bauer, B. J., Scherrenberg, R., Froehling, P., Joosten, J. and Amis, E. J. (1999). Fatty acid modified dendrimers in bulk and solution: single-chain neutron scattering from dendrimer core and fatty acid shell, *Macromolecules* **32**, 15, pp. 4983–4988, doi:10.1021/ma9901238.

Redfield, A. (1965). *Adv. Magn. Reson.*, **1**, pp. 1–32.

Reinitzer, F. (1888). Beiträge zur kenntniss des cholesterins, *Monatshefte für Chemie (Wien)* **9**, 1, pp. 421–441, doi:10.1007/BF01516710.

Richardson, R. M., Ponomarenko, S. A., Boiko, N. I. and Shibaev, V. P. (1999a). Liquid crystalline dendrimer of the fifth generation: from lamellar to columnar structure in thermotropic mesophases, *Liquid Crystals* **26**, 1, pp. 101–108, doi:10.1080/026782999205605.

Richardson, R. M., Whitehouse, I. J., Ponomarenko, S. A., Boiko, N. I. and Shibaev, V. P. (1999b). X-ray diffraction from liquid crystalline carbosilane dendrimers, *Molecular Crystals and Liquid Crystals* **330**, pp. 167–174, doi:10.1080/10587259908025588.

Rosen, B. M., Peterca, M., Huang, C. H., Zeng, X. B., Ungar, G. and Percec, V. (2010). Deconstruction as a strategy for the design of libraries of self-assembling dendrons, *Angewandte Chemie-international Edition* **49**, 39, pp. 7002–7005, doi:10.1002/anie.201002514.

Rueff, J. M., Barbera, J., Donnio, B., Guillon, D., Marcos, M. and Serrano, J. L. (2003). Lamellar to columnar mesophase evolution in a series of PAMAM liquid-crystalline codendrimers, *Macromolecules* **36**, 22, pp. 8368–8375, doi:10.1021/ma030223k.

Rueff, J. M., Barbera, J., Marcos, M., Omenat, A., Martin-Rapun, R., Donnio, B., Guillon, D. and Serrano, J. L. (2006). PAMAM- and DAB-derived dendromesogens: the plastic supermolecules, *Chemistry of Materials* **18**, 2, pp. 249–254, doi:10.1021/cm048105e.

Rutar, V., Vilfan, M., Blinc, R. and Bock, E. (1978). Deuteron spin-lattice relaxation mechanisms in partially deuterated nematic MBBA, *Molecular Physics* **35**, 3, pp. 721–728, doi:10.1080/00268977800100541.

Saez, I. M. and Goodby, J. W. (1999). Supermolecular liquid crystal dendrimers based on the octasilsesquioxane core, *Liquid Crystals* **26**, 7, pp. 1101–1105.

Saez, I. M. and Goodby, J. W. (2005). Supermolecular liquid crystals, *Journal of Materials Chemistry* **15**, 1, pp. 26–40, doi:10.1039/b413416h.

Saez, I. M., Goodby, J. W. and Richardson, R. M. (2001). A liquid-crystalline silsesquioxane dendrimer exhibiting chiral nematic and columnar mesophases, *Chemistry - A European Journal* **7**, 13, pp. 2758–2764, doi: 10.1002/1521-3765(20010702)7:13<2758.

Samulski, E. T. (2010). Meta-cybotaxis and nematic biaxiality, *Liquid Crystals* **37**, 6-7, pp. 669–678, doi:10.1080/02678292.2010.488938.

Sebastião, P. J., Ribeiro, A. C., Nguyen, H. T. and Noack, F. (1993). Proton NMR relaxation study of molecular motions in a liquid-crystal with a strong polar terminal group, *Zeitschrift Fur Naturforschung Section A - A Journal of Physical Sciences* **48**, 8-9, pp. 851–860.

Sebastião, P. J., Ribeiro, A. C., Nguyen, H. T. and Noack, F. (1995). Molecular-dynamics in a liquid-crystal with reentrant mesophases, *Journal de Physique II* **5**, 11, pp. 1707–1724.

Sebastião, P. J., Sousa, D., Ribeiro, A. C., Vilfan, M., Lahajnar, G., Seliger, J. and Žumer, S. (2005). Field-cycling NMR relaxometry of a liquid crystal above T-NI in mesoscopic confinement, *Physical Review E* **72**, 6, p. 061702, doi:10.1103/PhysRevE.72.061702.

Sebastião, P. J., Cruz, C. and Ribeiro, A. C. (2009). *Nuclear magnetic resonance spectroscopy of liquid crystals*, chap. 5. (World Scientific, Ronald Y. Dong ed.), pp. 129–167.

Sebastião, P. J., Godinho, M. H., Ribeiro, A. C., Guillon, D. and Vilfan, M. (1992). NMR-study of molecular-dynamics in a mixture of 2 polar liquid-crystals (CBOOA and DOBCA), *Liquid Crystals* **11**, 4, pp. 621–635.

Serrano, J. L., Marcos, M., Martin, R., Gonzalez, M. and Barbera, J. (2003). Chiral codendrimers derived from poly(propyleneimine) dendrimers (DAB), *Chemistry of Materials* **15**, 20, pp. 3866–3872, doi:10.1021/cm030250p.

Severing, K. and Saalwachter, K. (2004). Biaxial nematic phase in a thermotropic liquid-crystalline side-chain polymer, *Physical Review Letters* **92**, 12, p. 125501, doi:10.1103/PhysRevLett.92.125501.

Skoulios, A. and Guillon, D. (1988). Amphiphilic character and liquid crystallinity, *Molecular Crystals and Liquid Crystals* **165**, pp. 317–332, doi:10.1080/00268948808082205.

Slichter, C. P. (1992). *Principles of Magnetic Resonance*, Solid State Physics, Suppl. 1 (Springer-Verlag, Berlin/Germany).

Sluckin, T. J., Dunmur, D. A. and Stegemeyer, H. (2004). *Crystals That Flow: Classic Papers from the History of Liquid Crystals*, Liquid Crystals Book Series (Taylor & Francis), URL http://books.google.pt/books?id=iMEMAuxrhFcC.

Sousa, D. M., Fernandes, P. A. L., Marques, G. D., Ribeiro, A.C. and Sebastião, P. J. (2004). Novel pulsed switched power supply for a fast field cycling NMR spectrometer, *Solid State Nuclear Magnetic Resonance* **25**, pp. 160–166, doi: 10.1016/j.ssnmr.2003.03.026.

Sousa, D. M., Marques, G. D., Cascais, J. M. and Sebastião, P. J. (2010). Desktop fast-field cycling nuclear magnetic resonance relaxometer, *Solid State Nuclear Magnetic Resonance* **38**, pp. 36–43, doi: 10.1016/j.ssnmr.2010.07.001.

Stapf, S., Kimmich, R., R. and Niess, J. (1994). Microstructure of porous-media and field-cycling nuclear magnetic-relaxation spectroscopy, *Journal of Applied Physics* **75**, 1, pp. 529–537, doi:10.1063/1.355834.

Stebani, U. and Lattermann, G. (1995). Unconventional mesogens of hyper-branched amides and corresponding ammonium derivatives, *Advanced Materials* **7**, 6, pp. 578–581, doi:10.1002/adma.19950070617.

Straley, J. P. (1974). Ordered phases of a liquid of biaxial particles, *Physical Review A* **10**, 5, pp. 1881–1887, doi:10.1103/PhysRevA.10.1881.

Tinh, N. H., Destrade, C., Levelut, A. M. and Malthete, J. (1986). Biforked mesogens: a new type of thermotropic liquid-crystals, *Journal de Physique* **47**, 4, pp. 553–557.

Torrey, H. C. (1953). Nuclear spin relaxation by translational diffusion, *Physical Review* **92**, 4, pp. 962–969.

Trahasch, B., Frey, H., Lorenz, K. and Stuhn, B. (1999a). Dielectric relaxation in carbosilane dendrimers with cyanobiphenyl end groups, *Colloid and Polymer Science* **277**, 12, pp. 1186–1192, doi:10.1007/s003960050508.

Trahasch, B., Stuhn, B., Frey, H. and Lorenz, K. (1999b). Dielectric relaxation in carbosilane dendrimers with perfluorinated end groups, *Macromolecules* **32**, 6, pp. 1962–1966, doi:10.1021/ma981075e.

Tschierske, C. (1998). Non-conventional liquid crystals: the importance of micro-segregation for self-organisation, *Journal of Materials Chemistry* **8**, 7, pp. 1485–1508, doi:10.1039/a800946e.

Tsiourvas, D., Felekis, T., Sideratou, Z. and Paleos, C. M. (2004). Ionic liquid crystals derived from the protonation of poly(propylene imine) dendrimers with a cholesterol-based carboxylic acid, *Liquid Crystals* **31**, 5, pp. 739–744, doi:10.1080/02678290400001681618.

Tsiourvas, D., Stathopoulou, K., Sideratou, Z. and Paleos, C. M. (2002). Body-centered-cubic phases derived from n-dodecylurea functionalized poly(propylene imine) dendrimers, *Macromolecules* **35**, 5, pp. 1746–1750, doi:10.1021/ma010933v.

Tsvetkov, V. (1942). *Acta Physicochim. (USSR)* **16**, p. 132.

Ujiie, S., Yano, Y. and Mori, A. (2004). Liquid-crystalline branched polymers having ionic moieties, *Molecular Crystals and Liquid Crystals* **411**, pp. 1525–1531, doi:10.1080/15421400490436403.

Ukleja, P., Pirš, J. and Doane, J. W. (1976). Theory for spin-lattice relaxation in nematic liquid crystals, *Physical Review* **14**, p. 414.

Van-Quynh, A., Filip, D., Cruz, C., Sebastião, P. J., Ribeiro, A. C., Rueff, J. M., Marcos, M. and Serrano, J. L. (2005). NMR relaxation study of molecular dynamics in columnar and smectic phases of a PAMAM liquid-crystalline co-dendrimer, *European Physical Journal E* **18**, 2, pp. 149–158, doi:10.1140/epje/i2005-10036-4.

Van-Quynh, A., Filip, D., Cruz, C., Sebastião, P. J., Ribeiro, A. C., Rueff, J. M., Marcos, M. and Serrano, J. L. (2006). NMR relaxation study of molecular dynamics in the smectic A phase of PAMAM liquid crystalline dendrimers of generation 1 and 3, *Molecular Crystals and Liquid Crystals* **450**, pp. 391–401, doi:10.1080/15421400600588363.

Van-Quynh, A., Sebastião, P. J., Wilson, D. A. and Mehl, G. H. (2010). Detecting columnar deformations in a supermesogenic octapode by proton NMR relaxometry, *European Physical Journal E* **31**, 3, pp. 275–283, doi:10.1140/epje/i2010-10575-5.

Vanakaras, A. G. and Photinos, D. J. (2008). Thermotropic biaxial nematic liquid crystals: spontaneous or field stabilized? *Journal of Chemical Physics* **128**, 15, p. 154512, doi:10.1063/1.2897993.

Vilfan, M., Althoff, G., Vilfan, I. and Kothe, G. (2001). Nuclear-spin relaxation induced by shape fluctuations in membrane vesicles, *Physical Review E* **64**, 2, p. 022902, doi:10.1103/PhysRevE.64.022902.

Vilfan, M., Apih, T., Sebastião, P. J., Lahajnar, G. and Žumer, S. (2007). Liquid crystal 8CB in random porous glass: NMR relaxometry study of molecular diffusion and director fluctuations, *Physical Review E* **76**, 5, p. 051708, doi:10.1103/PhysRevE.76.051708.

Vold, R. and Vold, R. (1994). *The Molecular Dynamics of Liquid Crystals* (Kluwer Academic, the Netherlands).

Ward, B. B. and Baker Jr., J. R. (2008). *Dendrimer-Based Nanomedicine* (Pan Stanford, Singapore).

Woessner, D. E. (1962). Spin relaxation processes in a 2-proton system undergoing anisotropic reorientation, *Journal of Chemical Physics* **36**, 1, pp. 1–&, doi:10.1063/1.1732274.

Zavada, T. and Kimmich, R. (1999). Surface fractals probed by adsorbate spin-lattice relaxation dispersion, *Physical Review E* **59**, 5, pp. 5848–5854, doi:10.1103/PhysRevE.59.5848.

Żavada, Z. and Kimmich, R. (1998). The anomalous adsorbate dynamics at surfaces in porous media studied by nuclear magnetic resonance methods. the orientational structure factor and levy walks, *Journal of Chemical Physics* **109**, 16, pp. 6929–6939.

Zhu, X. M., Vinokur, R. A., Ponomarenko, S. A., Rebrov, E. A., Muzafarov, A. M., Boiko, N. I. and Shibaev, V. P. (2000). Synthesis of new carbosilane ferroelectric liquid-crystalline dendrimers, *Polymer Science Series A* **42**, 12, pp. 1263–1271.

Zhu, X. M., Boiko, N. I., Rebrov, E. A., Muzafarov, A. M., Kozlovsky, M. V., Richardson, R. M. and Shibaev, V. P. (2001). Carbosilane liquid crystalline dendrimers with terminal chiral mesogenic groups: structure and properties, *Liquid Crystals* **28**, 8, pp. 1259–1268.

Zimmermann, H., Poupko, R., Luz, Z. and Billard, J. (1985). Pyramidic mesophases, *Zeitschrift Fur Naturforschung Section A - A Journal of Physical Sciences* **40**, 2, pp. 149–160.

Žumer, S. and Vilfan, M. (1978). Theory of nuclear-spin relaxation by translational self-diffusion liquid-crystals: nematic phase, *Physical Review A* **17**, 1, pp. 424–433, doi:10.1103/PhysRevA.17.424.

Žumer, S. and Vilfan, M. (1978). Translational diffusion in smectic liquid-crystals and its effect on nuclear-spin relaxation, *Journal of Molecular Structure* **46**, MAY, pp. 475–480.

Žumer, S. and Vilfan, M. (1980). Theory of nuclear-spin relaxation by translational self-diffusion in liquid-crystals: smectic-A phase, *Physical Review* **21**, 2, pp. 672–680.

Žumer, S. and Vilfan, M. (1981). Disc-like liquid-crystals: molecular motions and nuclear magnetic-relaxation, *Molecular Crystals and Liquid Crystals* **70**, 1-4, pp. 1317–1334.

Žumer, S. and Vilfan, M. (1983). Nuclear-spin relaxation due to translational diffusion in a hexatic-b and crystalline-b phase, *Physical Review A* **28**, 5, pp. 3070–3079.

Zupančič, I., Pirš, J., Luzar, M., Blinc, R. and Doane, J. W. (1974). Anisotropy of self-diffusion tensor in nematic MBBA, *Solid State Communications* **15**, 2, pp. 227–229, doi:10.1016/0038-1098(74)90746-7.

Zupančič, I., Zagar, V., Rozmarin, M., Levstik, I., Kogovsek, F. and Blinc, R. (1976). Angular-dependence of proton spin-spin and spin-lattice relaxation in nematic MBBA-EBBA mixtures, *Solid State Communications* **18**, 11–1, pp. 1591–1593, doi:10.1016/0038-1098(76)90400-2.

Index